D1102300

Collins

how can i stop climate change?

Helen Burley and Chris Haslam

Collins

Friends of the Earth

www.foe.co.uk/living

HarperCollins Publishers
77-85 Fulham Palace Road
London W6 8JB

Collins website: **www.collins.co.uk**
Friends of the Earth website: **www.foe.co.uk**

Collins is a registered trademark of HarperCollins Publishers Ltd.

First published in 2008

A catalogue record for this book is available from the British Library.

ISBN 978 0 00 726163 5

The paper used for the text pages is made from de-inked 100% post consumer waste.

Collins uses papers that are natural, renewable and recyclable products made from wood grown in sustainable forests. The manufacturing processes conform to the environmental regulations of the country of origin.

Editor: Adam Bradbury
Designer: Martin Brown
Picture research: Calliste Lelliott, Amelia Collins
Indexer: Ben Murphy
Project Manager HarperCollins: Julia Koppitz
Commissioning Editor HarperCollins: Myles Archibald

Printed and bound in Italy

acknowledgements

Additional writing and research by Dr Martin Cullen, Valentine Karsenty and Rachel Skerry

Friends of the Earth would like to thank many people who helped create this book, especially:
Franny Armstrong, Nicola Baird, Jenny Bates, Raoul Bhambral, Theo Bird, Mike Birkin, Tony Bosworth, Heather Buttivant, Kirtana Chandrasekaran, Mike Childs, Giles Cory, Stuart Croft, Neil Crumpton, Pooran Desai, Richard Dyer, Katie Elliott, Lizzie Gillett, Hannah Griffiths, Barry Grommett, Roger Higman, Vicki Hird, Kurt Jackson, Tim Jenkins, Tom Kenward, Lou Krzan, Sasha Lean-Vercoe, Calliste Lelliott, Charlotte Leyburn, Stephanie Long, Naomi Luhde-Thompson, Mark Lynas, Rita Marcangelo, Ed Matthew, Martin Normanton, Alex Phillips, Nick Rau, Steve Shaw, Graeme Sherriff, Mary Taylor, Dave Timms, Tony Wainwright, Penny Walker, Adeela Warley, Michael Warhurst, Robin Webster, Claire Weir, Martyn Williams, Bryony Worthington, Paul de Zylva.

Special thanks to Martin Cullen for his help with research, fact-checking and editing.

And, to the many people named in the book whose stories help show that together we can beat climate change, thank you.

Many organisations have also provided information; contact details are listed at the end of this book.

p.16: data from Brohan *et al.*, 2006. *J. Geophysical Research* **111**, D12106, doi:10.1029/2005JD006548). For more, go to www.cru.uea.ac.uk

p.52: Dr Keeling's measurements at Mauna Loa are being continued by the Scripps Institute of Oceanography and by the National Oceanic and Atmospheric Administration. For more, go to www.esrl.noaa.gov/gmd/ccgg/trends/co2_data_mlo.html

picture acknowledgements

The publishers thank the following for their kind permission to reproduce the photographs in this book:

Photo libraries:
Antoine Gyori/AGP/Corbis
Artiga Photo/Corbis
Ashley Cooper/Corbis
Bloomimage/Corbis
Branko Miokovic/istock
Car Culture/Corbis
Chris Mattison/FLPA
Chris Mole
Christian Denes
Daniel Munoz/Reuters/Corbis
Darren Greenwood/Design Pics/Corbis
Darryl Bush/San Francisco Chronicle/Corbis
David Gray/Corbis
David Madison/Corbis
David Papazian/Beateworks/Corbis
Department for Environment, Food and Rural Affairs
Dieter Telemans/Panos
Digital Vision
Earth-vision.biz
ExCel Centre, London
Findlay Kember/AFP/Getty
Gavriel Jecan/Corbis
George B. Diebold/Corbis
Gideon Mendel/Corbis
G.M.B.Akash
Grzegorz Lepiarz/istock
Helen King/Corbis
Igor Kostin/Sygma/Corbis
istock
James Leynse/Corbis
Jan Martin Will/Shutterstock
Jason Hawkes/Corbis
Jason Lee/Reuters
jmarley12/Flickr
Joe Cornish/Arcaid/Corbis
Joe Cornish/NTPL
Jolanda Cats & Hans Withoos/zefa/Corbis
Kate Mitchell/zefra/Corbis
Larry Williams/Corbis
Living Villages
Mark E. Gibson/Corbis
Mark Evans/istock
Matt Rainey/Star Ledger/Corbis
Max Hogg
Medio Images/Corbis
Micha Rosenwirth/istock
Natalie Behring/Panos
Nicolas Pousthomis/sub.coop
Oman Mirzaie/istock
Orangutan Foundation
Paul Colangelo/Corbis
Pines Calyx
Punit Paranjpe/Reuters
Quiet Revolution
Randy Faris/Corbis
Roy Botterell/zefa/Corbis
Roy Rainford/Robert Harding World Imagery/Corbis
Sandra Ivany/Brand X/Corbis
Sarah Lee/Guardian News Ltd
Sean Justice/Corbis
Sigrid Dauth/Alamy
Solar Century
Spannerfilms
Staffan Widstrand/Corbis
Steven Vidler/Eurasia Press/Corbis
Tay Rees/ Getty
Tim Pannell/Corbis
Tom Brakefield/Corbis
Walter Rawlings/Robert Harding World Imagery/Corbis
Woking Borough Council

Friends of the Earth photographers:
Adam Bradbury
Alex Phillips
Amelia Collins
Aulia Erlangga
Balthazar Serreau
Chris Haslam
Georgina Cranston
Ian Lander
Naomi Luhde-Thompson
Paul Glendell
Tim Normanton

contents

foreword

" I was recently at the launch of the new high-speed rail link from London to Paris. Passengers boarded at eleven in the morning and arrived right in the heart of Paris just 180 minutes later. As the train slipped through the autumn countryside of southern England and northern France people talked face to face, read the morning papers and ate and drank in comfort and quiet. The people on Eurostar's *Tread Lightly* might not have known it but they were also doing the planet a favour. Each passenger going from London to Paris by train is creating just a tenth of the climate-changing emissions of someone making the journey by aeroplane.

For me that day was a moment of inspiration, a reminder of how humanity's technical brilliance, plus a dose of common sense, can take us - literally - to places we once thought impossible.

This book is about a better life. It's about turning the challenge of climate change into a way forward for us all, and coming out on the other side healthier, fitter and happier. And because we need to do something now, rather than in 40 years' time, it's about making that journey at high speed.

Our starting point is the science - what's happening to our world and why. The story has been brilliantly told by others, notably by the Nobel prize-winning scientists who make up the Intergovernmental Panel on Climate Change and former US vice-president Al Gore. We explain some key facts - about what's going on, and what's likely to happen if we let it (Chapters 1 to 3) - before moving on to the big question: what are we going to do about it?

Imagine a home that you don't need to heat because the sun does it for you. Imagine your fridge running on the never-ending power of the tides that beat at Britain's shores. Imagine buses so reliable and clean you actually want to leave the car at home. Imagine the end of fuel poverty, better food, fewer traffic jams and happier neighbourhoods. Imagine a new economic wave, new jobs in new industries supplying whole new markets. All this happens to be possible because the things that we need to do to stop climate change (read about them in Chapters 4 to 9) will also improve our quality of life.

The solutions are out there, but who's going to make them happen? From fitting low-energy light bulbs, to flying less and recycling more, there are many things each of us can do (lots of practical information in Chapters 7 to 9). Among the green self-help manuals around encouraging each of us to do our bit to take the stress off the planet, few make this key point: that even if each of us were to deal with the environmental impact of our home, our travel and our shopping, it would still not be enough. Yes I can turn off the lights but I don't run the polluting power station that makes the electricity. I can use my car less but why should I, if some gas guzzler is allowed to take up my space on the road? I can insulate my loft and even install solar panels on my house but what about my neighbour who can't afford it, and what about my kids' school, which uses far more energy than my family does?

Most climate-changing pollution is outside my direct control. Greenhouse gases are just part of how things work – like the way food is produced, public institutions are run, our roads are built. I can directly control only about a third of my own tiny share of Britain's carbon footprint. Someone has got to change the way things work on a big scale - and fast. In fact the science implies that the UK should cut its emissions of the main greenhouse gas, carbon dioxide, by at least 80 per cent by the middle of this century. When you think about how dependent we are now on oil, gas and coal, this looks like a huge task. Who's going to do it?

As consumers we have a certain amount of muscle – and we should use it. Nagging our shops and services for

greener stuff helps to push them in the right direction: being an ethical consumer also tells the politicians that we're ready for change. But we're not just shoppers. Each of us is part of a family, neighbourhood, or town. We can talk to our friends and family, get things done in our clubs, societies, schools, unions, and associations to reduce their carbon footprint.

Most important, we have rights and a voice as citizens. We can tell our elected representatives – from the town hall to parliament – that we want a greener life. They should get on and make it easier for all of us to do the low-carbon thing. Governments can ban cars with poor fuel consumption, set tough standards for products and provide alternatives to flying. Governments can structure markets to make companies innovate: obliging local authorities to specify clean green energy for all new homes, for example, would create a huge business opportunity and, as other countries have shown, do a lot to cut emissions.

But this scale of change will only happen if enough of us ask for it. When I joined the staff of Friends of the Earth in 1990 one of the first things I did was to promote our findings on how halting deforestation would be good for the climate. Since then we have notched up a series of wins for the environment. We have put green issues firmly on the political agenda, sought broad changes to the way our economy works and persuaded politicians to take real steps to protect our communities and future generations. And we've always done it with strong public backing: the reason you have doorstep recycling is because thousands of people like you asked for it.

At about the time this book will first appear in the shops a new Climate Change Act will be passing into UK law. This will be the world's first legally binding national framework for long-term reductions in greenhouse gas emissions. If the government has done its job properly the law will set a target for cutting carbon dioxide emissions by 80 per cent by 2050, including emissions from aviation and shipping. Friends of the Earth's Big Ask campaign and the voices of hundreds of thousands of people have been crucial in securing this breakthrough. It shows how powerful we are when we act together. In all my years working on the environment I have never been so excited about the changes that we're about to see – changes that people like you have made possible. There's a lot do, but we are making real progress.

The book in your hands is one of the next steps. It is the perfect companion for making the most of the opportunity created by the new Climate Change Act. The solutions are out there. Get on board and make them live.

None of us can deal with climate change alone - we have to act together. How can I stop climate change? I can't. But *we* can.

Tony Juniper

Tony Juniper, Executive Director, Friends of the Earth, 2002–08.

FRIENDS OF THE EARTH

Executive director, Tony Juniper

"The honest answer to the question 'How can I stop climate change?' is 'I can't. But *we* can.'"

the climate is changing around us

Flood, drought, storm, heatwave. What's happening to the weather?

chapter 1

a warmer world

Even without the scientists telling us, we can all see a new pattern emerging: freaky weather, unusual temperatures, ice sheets crashing into the sea, migrating birds turning up ahead of schedule, plants flowering early...

From the UK and Europe to Australia and the Americas, this chapter offers some snapshots of the way our planet is changing around – and because of – us.

As surely as the Wimbledon tennis tournament brings rain, fresh strawberries say lazy summer sunshine. But in 2007 you were more likely to be eating strawberries at Easter than in June. A bizarrely hot spring meant they were ripe and ready to pick by April. Too early for Wimbledon. Too early for the students who normally pick them. And too early for the strawberry farmers who employ the students.

The last time April in England was as warm as in 2007 Abraham Lincoln was still alive and Lewis Carroll was polishing

The graph below shows the global temperature for each year from 1850 to 2006. Climate experts use 14 °C as a reference point. The wavy line reveals the unmistakeable trend – that the world has been getting hotter over the past 150 years.

Global warming: the hard data

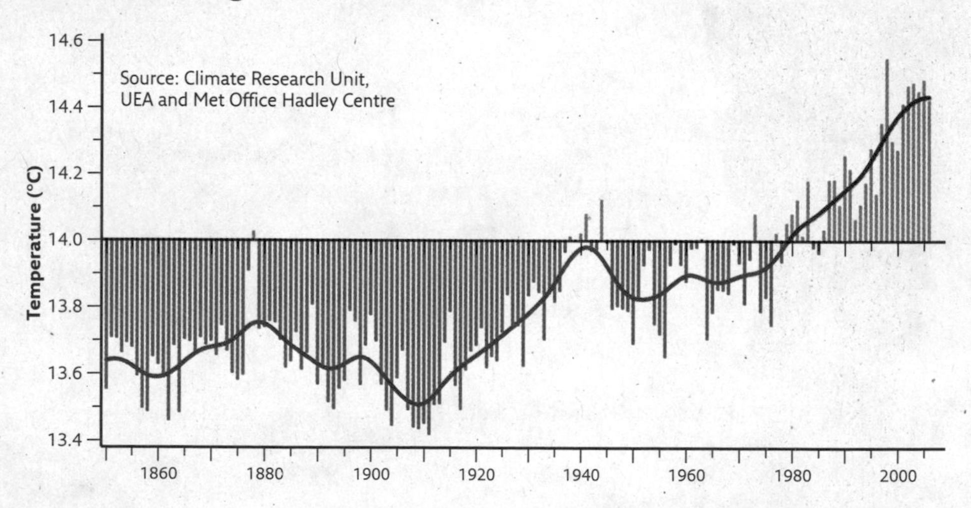

off *Alice in Wonderland*. It followed an unusually mild January, February and March. Frogspawn thrived in garden ponds in February, and insects like peacock butterflies, bumblebees and ladybirds appeared earlier than usual. Then came what looked like being the wettest summer in the UK since the outbreak of the First World War.

As England basked in its April sunshine Spain was doused in unseasonal rain, with disastrous consequences as salad crops were flooded, leading to soaring tomato prices. And then, while the UK saw flooding, Europe flipped into yet another heat wave.

Of course our weather is changeable – it always has been. One hot day doesn't make a summer, and a few record temperatures don't necessarily prove that the world's climate is in the throes of fundamental change. But average temperatures are definitely rising – as worldwide records clearly show. And we are now seeing the very kinds of changes to weather patterns and the Earth's natural systems, habitats and wildlife that scientists have for years been predicting will flow from global warming.

weird UK weather facts

fact Brogdale in Kent holds the record for the hottest-ever recorded temperature in the UK: 38.5 °C on 10 August 2003.

fact In June 2004 the wind blew across Northern Ireland at 126 miles per hour, leaving 8,000 homes without electricity.

fact Summer 2003 broke UK records for the hottest weather – with fatal results for more than 2,000 people. In mainland Europe the death toll topped 35,000.

fact July 2006 was the hottest month on record in the UK.

fact In April 2007 the average monthly temperature in Central England reached 11.6 °C, the highest for the month since records began in 1659.

three need-to-know terms

Greenhouse effect: How the Earth keeps warm. Greenhouse gases in the atmosphere trap heat from the sun, raising the temperature down here. Without greenhouse gases, our planet would be too cold to support most life.

Global warming: The way the average temperature on the Earth's surface is increasing. Scientific evidence clearly shows an increase of greenhouse gases is largely to blame.

Man-made warming: Greenhouse gases – the main ones are carbon dioxide, nitrous oxide, methane and water vapour – are naturally present in the Earth's atmosphere. The amount of greenhouse gases has remained relatively stable for thousands of years, but in the past century they have increased sharply. This is largely as a result of burning fossil fuels such as oil, coal and gas, which produces carbon dioxide.

five myths about climate change

'The Earth is so big that human beings cannot possibly affect it.'
Fact: While the Earth has for millions of years been capable of balancing emissions of carbon dioxide – with plants, soil and sea soaking up the carbon dioxide emitted by nature – the extra carbon dioxide that humans have put into the system has upset the balance. The atmosphere is surprisingly thin: if the Earth were the size of a football the atmosphere would be no thicker than a coat of varnish.

'Scientists don't agree that climate change is caused by humans.'
Fact: A 2004 study of 900 published papers on climate change found they all agreed that climate change was happening and was caused by human activity. There are

some sceptics – but the fiercest arguments today tend to focus on the scale of the problem and the best way to tackle it rather than whether or not climate change is happening at all.

'We have to choose between the economy and saving the planet.'
Fact: There is no such choice – in the long term, without a stable climate there will be no economy to save: the impacts on society of runaway climate change would be catastrophic. Keeping climate change in check will involve developing new technologies and industries and this will have economic benefits; and although investing in such opportunities has a cost, the cost of not dealing with climate change will be higher.

'It's too late to do anything about climate change.'
Fact: Some climate change is already happening. But it will get much worse if we let it. The more greenhouse gases we allow into the atmosphere, the greater the effect on our climate will be. Scientists warn that we are approaching tipping points that will trigger ever-more rapid changes to the climate, so the sooner we act, the better.

'Bring it on. The UK's climate will be much warmer – life here will be better.'
Fact: Britain is likely to be warmer and some people may welcome that. But we're also likely to see heavier rain, flooding and storms, as well as heat waves and drought in the summer. Water shortages have already hit the south-east of England. Impacts elsewhere will come sooner and harder, and the UK is likely to feel the knock-on effects – as places become inhospitable, crops fail and huge numbers of people are uprooted.

climate change is already affecting us

As climate systems adapt to changes in temperature, winds and ocean currents alter. This is leading to unpredictable and more extreme weather: more intense periods of heavy rain, more storms, and, in some regions, extreme drought. Climate change is having other knock-on effects. Rising temperatures are melting snow and ice all over the world: glaciers are retreating, Arctic ice sheets are shrinking and Antarctic ice shelves have cracked and fallen into the sea. Melting glaciers, combined with the fact that warmer water expands, are causing rising sea levels. Sea levels rose by 17 cm during the 20th century.

Many people's lives are already being altered by climate change. Some are adapting, but others face real hardship.

Let's look at some of the specific ways that climate change is affecting our planet.

rain and storms

Four people died and thousands had to evacuate their homes when torrential rain hit England in June 2007, causing flash floods, chaos to travel and billions of pounds worth of damage. A month later the heavy rains returned, leaving a swathe of central England under water. The Severn and Avon rivers flooded thousands of homes; the Thames burst its banks in Oxford and flowed into nearby villages. Roads were closed and train services failed. Between May and July twice as much rain as normal fell, breaking records going back to 1766.

Although unusual, the summer floods of 2007 were not entirely without precedent. Back in 1920 the River Lud in Lincolnshire burst its banks, washing away bridges, houses and

A RIVER RUNS THROUGH IT:

Tewkesbury in Gloucestershire got a taste of a future with climate change in the summer of 2007 – among the wettest ever for many communities across Britain.

cars in the small market town of Louth; 23 people died. Homes and businesses in Louth were under water again in 2007.

It's impossible to pin the blame for one weather event on climate change. But scientists say our changing climate is making such extremes more likely.

Yet the UK experience to date has been relatively mild. In August 2002 floods swept central Europe: Germany had a record-breaking 35 cm of rain in 24 hours and in Dresden, water levels of the River Elbe reached their highest since records began in 1275. In 2005 more than 65 cm of rain fell in Mumbai, India, in 24 hours – the country's heaviest ever recorded rainfall. In China torrential summer rains in 2007 affected an estimated 100 million people and hundreds died in flash flooding and landslides. In South Asia millions fled their homes and thousands died after heavy monsoon rain caused flooding in parts of Northern India, Nepal and Bangladesh. In Assam 90 cm of rain fell in 20 days, flooding farmland.

hurricanes, cyclones and monsoons

Hurricanes over the Atlantic during the summer are nothing new. But the 2005 season made many people think about climate change again. Hurricane Katrina, which tore through New Orleans in August that year, was not the most powerful of the storms, but it did the most damage. Nearly 2,000 people died and many thousands lost their homes – and this in the world's richest country.

More recently hundreds died in Pakistan, Afghanistan and India in 2007 when Cyclone Yemyin hit the region. The disaster was made worse by heavy monsoon rains bringing severe flooding.

melting glaciers and ice sheets

Rising temperatures are damaging the major ice sheets at the Poles. In Greenland, which is warming faster than other parts of the world, vast glacial rivers are cutting through the ice. In the summer of 2007 sea ice shrank to a record low, with an area twice the size of Britain disappearing in the space of a week. Scientists predict that by 2030 the Arctic could be ice-free in the summer.

Generations of Inuit in the Arctic Circle have depended on hunting and fishing to feed their families, but changes in the sea ice make navigation difficult and hunting on thin ice perilous. Experienced hunters have died where the ice has given

THAW POINT:
Argentina's Perito Moreno Glacier is one of the few that are not retreating because of global warming. But for how much longer?

way. Aqqaluk Lynge, President of the Inuit Circumpolar Council, Greenland, says, 'Climate change in the Arctic is not just an environmental issue with unwelcome economic consequences. It is a matter of livelihood, food, and individual and cultural survival. The Arctic is not wilderness or a frontier, it is our home and homeland.'

Further south, in the semi-frozen areas of Alaska and Siberia, thawing of the once-permanently frozen ground is turning solid foundations for buildings and roads into soft bog. Houses are sinking as the land subsides.

Warmer weather is also causing problems for Europe's snowy regions and the people who live and work there. Alpine glaciers are in retreat and snow is arriving later and melting earlier in the year. Alpine glaciers shrank by 10 per cent in 2003 alone. Germany's last glacier, the Zugspitze, which provides drinking water for the Rhine Valley, is melting so rapidly that people in the area have tried to protect it from the sun by covering it with canvas.

a lifetime of change

Fernand Pareau has been an Alpine mountain guide in Chamonix, France, for more than 50 years. Aged 81 in 2007, he has witnessed huge changes, including more tourists and more traffic in the mountains.

He says the natural environment has altered, with a shift in the seasons, and changes in snowfall. Whereas there were once four distinct seasons, now winter comes late and summer comes sooner. Birds like magpies – previously not seen at such altitudes – are now common.

'It's not beautiful anymore, because when you're in Chamonix, you look at the Bosson Glacier... it's black, so it's not beautiful. You see the waterfalls there but it's not white anymore, it's ugly,' says Fernand. 'I think that for the future, if global warming continues like this, people will come to the mountains much less.'

melting in the Himalaya

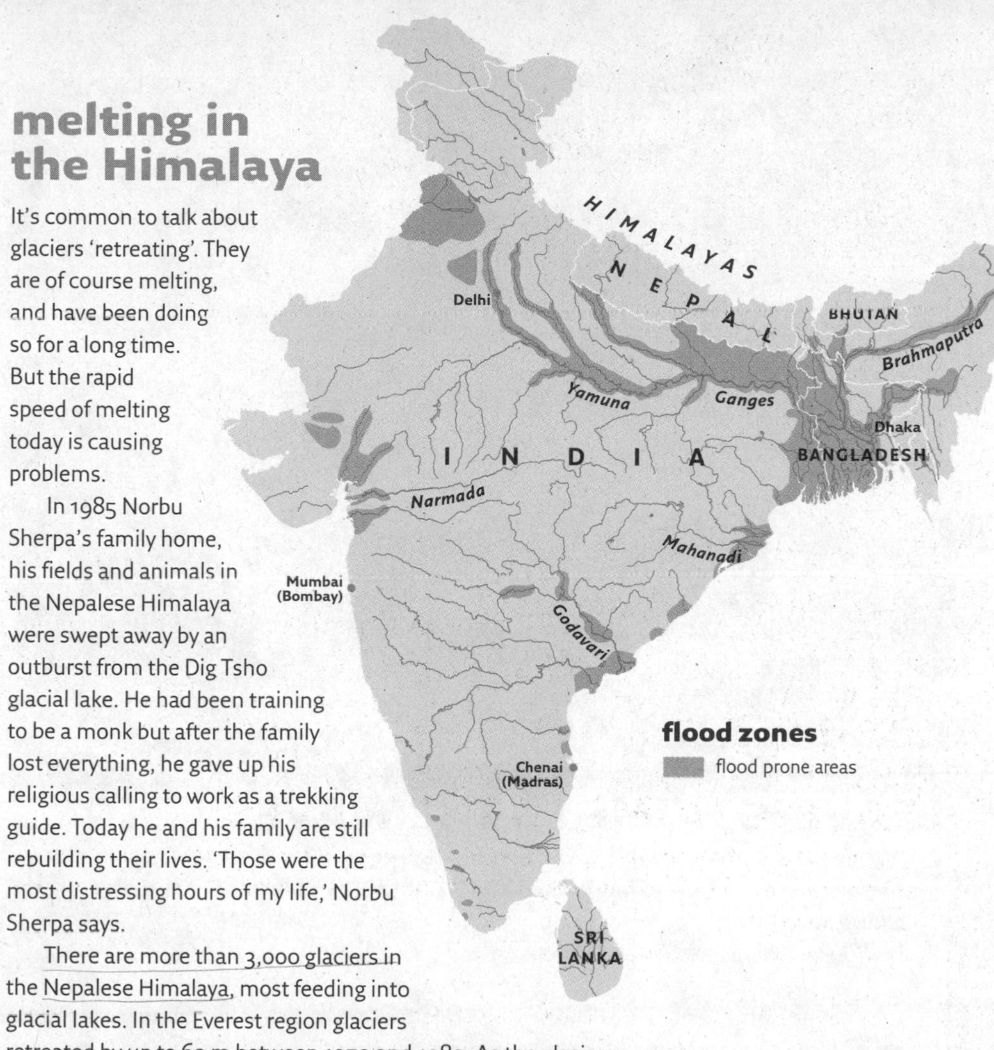

It's common to talk about glaciers 'retreating'. They are of course melting, and have been doing so for a long time. But the rapid speed of melting today is causing problems.

In 1985 Norbu Sherpa's family home, his fields and animals in the Nepalese Himalaya were swept away by an outburst from the Dig Tsho glacial lake. He had been training to be a monk but after the family lost everything, he gave up his religious calling to work as a trekking guide. Today he and his family are still rebuilding their lives. 'Those were the most distressing hours of my life,' Norbu Sherpa says.

There are more than 3,000 glaciers in the Nepalese Himalaya, most feeding into glacial lakes. In the Everest region glaciers retreated by up to 60 m between 1970 and 1989. As the glaciers retreat, water levels in the lakes rise. A study for the United Nations Environment Programme warned that more than 20 lakes in the Himalayas are potentially dangerous because of climate change – at risk of bursting their banks.

The retreat of Himalayan glaciers also poses a threat to the millions of people in India and across Asia who rely on snow and glacial melt for their water; scientists think that in 20-30 years these supplies will be greatly reduced.

LEAVING HOME TO THE WAVES:

'When I was a small boy this island was big. As I grew older the island got smaller As you can see the island has broken and it's now in two pieces. The sea is eating the island away.'

Jacob Tsomi, chief of the Dog clan, Carteret Islands

The Carteret Islands – a small horseshoe of atolls in the Pacific Ocean – form part of Papua New Guinea. Their once-fertile vegetable gardens are now all poisoned by salt. Lying less than 1.5 metres above sea level, the islands are frequently flooded by tidal surges. Bernard Tunim, who lives on Piul Island, describes the effects of a tidal surge: 'We planted banana, taro, even tapioca, cassava and other fruit trees. But just last month we had these high tides and it swept the whole area. We had waves coming from three sides, and so this place was flooded with sea water. All that we had from banana to taro and cassava just dried up. But we don't want to move because we know that this is our own land.'

Bernard says the islanders are angry at the prospect of having to leave their home – victims of a problem they are not responsible for. 'We believe that these islands are ours and our future generation should not go away from this island.'

rising rivers

Seasonal flooding is a way of life for people on the flood plains of Bangladesh; but recent years have seen a combination of heavy monsoons, deforestation and faster glacial melting swelling the rivers. Andean villagers in Peru face a similar threat: it's estimated that some 30,000 people have died as a result of sudden glacial floods. Disappearance of the ice also poses a threat to the water supply for the 7 million people who live in Lima.

rising seas

The coastlines of Vietnam and Bangladesh, small islands in the Pacific and Caribbean, and large coastal cities such as Tokyo, New York, Cairo and London are all threatened by rising seas.

Flood defences such as the London Thames Barrier can provide some protection – at a cost. The £535 million barrier was designed to offer protection against highly infrequent but dangerous storm surges. It had been used 103 times by 2007.

The impacts of higher seas are well known: high tides and storm surges come further and further inland, damaging property, washing away roads, contaminating fresh water and making it difficult for anything to grow. The low-lying Pacific island Tuvalu now depends on imported food and its people are seeking refugee status in neighbouring New Zealand as the sea takes over their homes.

Thames Barrier, 1983–2007

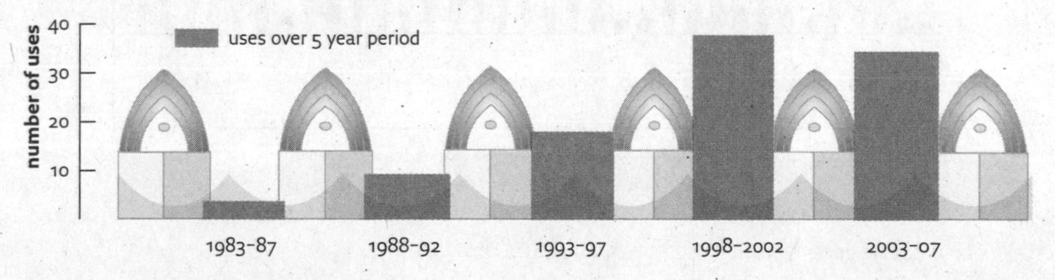

wildfires

Hot, dry weather increases the risk of wildfire. In recent years wildfires have rampaged through southern Europe, Australia, California and the Amazon.

Wildfires can improve conditions for vegetation and encourage wildlife. But with dry weather and strong winds they can also burn out of control. As temperatures rise, northern California is predicted to see wildfires increase by up to 90 per cent by the end of the century. In 2007 Greece experienced some of the worst wildfires ever recorded in the country, destroying forests around Olympia and scorching stones that have been there for 2,500 years. Sixty-five people died in a spate of fires over a ten-day period.

Britain's moorlands, particularly peat bogs in the Peak District, are vulnerable. In the summer of 2003 satellite images showed clouds of smoke drifting from the moors over the Irish Sea. Aircraft at Manchester Airport were unable to land and drivers kept their headlights on during the day for several days.

Scientists pin the blame for an increase in wildfires in the western United States on rising temperatures (below).

Forest in flames

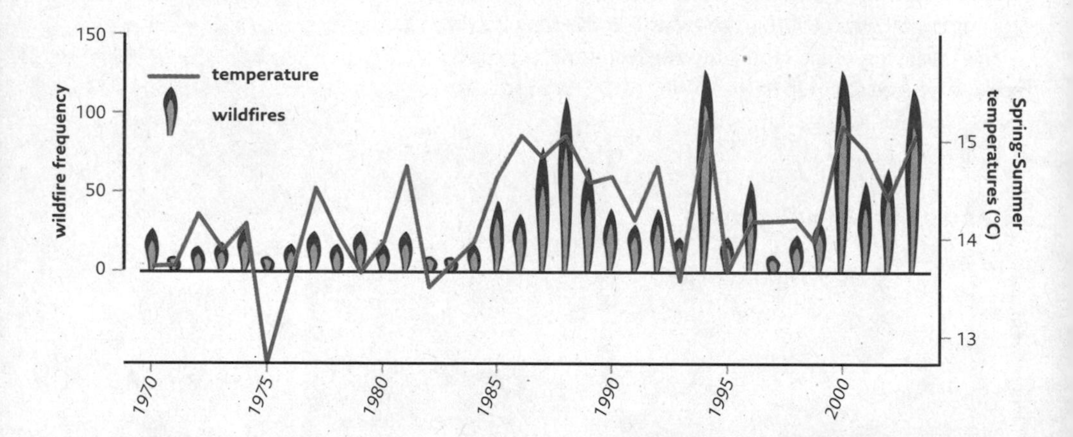

where we're feeling the heat

Climate scientists predict that with summers becoming hotter and drier in some regions, there will be more cases of extreme drought. Here's how different parts of the world are already being affected.

By 2100 – within the lifetimes of our grandchildren – half of the world's land could see reduced rainfall, causing drought.

record-breaking temperatures in the UK

Temperatures in central England have increased on average by 1°C since 1960, and individual months are getting warmer year by year.

People are feeling it. The heat wave of summer 2003 killed more than 2,000 people in the UK, and in 2006 a heat wave led to a shortage of grazing land for dairy herds, pushing up the costs of milk production. A 2006 survey of farmers found 60 per cent claiming that they were already feeling the effects of climate change. In some parts of southern England farmers are introducing crops once thought more appropriate to more exotic climes, including apricots, sunflowers and maize. Olives are thriving in Devon and English wines are gaining fans as growing conditions become more suitable. But not all fruits are benefiting – milder winters are not very good for blackcurrants, for example, which need cold weather to bear fruit. And drier summers mean that more and more fruit farmers are having to irrigate their crops.

heat waves in Europe

The past decade has seen chaotic weather across Europe. A heat wave in 2003 contributed to the death of an estimated 35,000 people across the continent, prompting experts to call it one of the deadliest climate-related disasters in Western history. Financially the damage was estimated at more than 13 billion euros. Harvests were badly hit and in France six power stations had to close because of low water levels in the rivers used for cooling systems.

And it appears 2003 was not an exception: temperatures again hit record highs of 46°C in south-east Europe in 2007, contributing to the deaths of more than 500 people in Hungary. According to climate experts, summers like 2003 are likely to become more and more common.

CURRANT CONCERN:

Fruits such as apple, strawberry, and blackcurrant need a sustained cold period to flower and fruit normally. But winters are becoming progressively milder – worrying times for the UK's £230 million-a-year fruit industry.

English wine – treading new ground

Source: Indicators of Climate Change in the UK (ICCUK)

The expansion of English and Welsh vineyards, although not attributable to climate change alone, shows the impact of warmer summers, market forces and wine-makers' expectations of global warming within our lifetime.

have potatoes had their chips?

Walter Simon has been growing potatoes in Pembrokeshire, Wales, for more than 20 years. Farming has altered in that time, he says, as a result of changes in the weather. Planting happens earlier, and the harvest comes earlier too. Walter concedes that milder winters have made life on the farm easier in some ways – outside pipes no longer need lagging and the sheds where he lays out the potatoes for seed no longer have to be proofed against draughts. Now his main worry is making sure the sheds get enough ventilation to keep them cool.

Walter grows early potatoes and relies on irrigation to water his crop, but he worries about growing seasons extending elsewhere in the country. An earlier harvest in the east of England, for example, could push him out of the market. In 20 years' time, he says, they probably won't be growing potatoes in Pembrokeshire. 'We'll be growing apricots or something,' he suggests. 'But it is the speed of the change that concerns people. It's not been as gradual an evolution as things may have been in the past.'

down under gets a roasting

In the space of just five years, 2002-2006, Australia suffered three of its worst droughts on record. 'A frightening glimpse of the future with global warming' was how South Australia's Premier, Mike Rann, described the 2006 drought.

In South West Australia annual total rainfall has declined by some 15-20 per cent in the past 30 years. As elsewhere an early casualty is farming. Harvests have been failing completely or drastically reduced. Rice production has plunged by more than 90 per cent during the past decade. Irrigated crops such as citrus and vines are particularly vulnerable as water levels decline (grape production fell by nearly a third in 2007).

The high temperatures are having some more bizarre consequences. Players at the Australian Open tennis tournament in 2007 had to abandon the outside courts during the day because it was too hot. They restarted in the evening and went on into the night – one match ended at 3 am.

BREAK POINT:

Global warming affects our world from top to bottom, including sporting events such as the Australian Open, where 40 °C+ daytime temperatures forced night time play in 2006.

farming in a drier world

Farmer Alan Brown has survived ten years of below-average rainfall in New South Wales, Australia. In 2006 he had his worst year to date when, for the second time, his harvest completely failed.

An established farmer with 900 hectares of land and a mixture of sheep, cattle and winter crops, Alan says, 'Everything I have revolves around the value of my land – and if my land is not producing, it isn't worth anything.' With less and less grazing available, Alan has taken to hand-feeding the animals for up to eight months of the year (he would normally do this for only three).

Winter rains are crucial in New South Wales. Moisture does not remain in the soil for long, and when winter rains fail, the outlook is dire, particularly for farmers who depend on irrigation. The financial effects are being felt
in the wider community. Villages that once supported several shops and services are down to just one general store. Parents are finding it cheaper to move their children to cities for schooling – thus splitting the family. Alan worries that Australia is witnessing a prolonged drying, symptomatic of climate change. 'If we are going to
survive in a drier environment, we need plants that can survive with less water. It's not something that we have bred for in the past.'

drying up in the Amazon

Covering an area of more than 2 million square miles, and home to a third of all animal species, the Amazon rainforest has been described as the Earth's lungs. It absorbs vast quantities of carbon dioxide from the atmosphere, releasing oxygen and playing a crucial role in keeping our climate on an even keel.

But as our planet warms, this great natural resource is at risk from long periods of dry weather. In 2005 water in the Amazon River was so low that sections were impassable by boat. The Brazilian army was called in to distribute water and food, and big ships were left stranded. In one state alone, fire laid waste to 100,000 hectares of forest – an area two-thirds the size of Greater London. Scientists estimate that burning in the Amazon adds some 370 million tons of greenhouse gases into the atmosphere every year.

did you know?

Some scientists say climate change is affecting the frequency of El Niño – the occasional reversal of the weather over the Pacific region. This phenomenon has a huge influence on weather systems around the world. During El Niño flooding becomes more likely on the Pacific coast of the Americas and cold water fish supplies disappear.

the spreading desert

Dust storms are common in China when the wind blows. In northern China once-fertile land is being destroyed as the heat and lack of water kill vegetation. The dry soil quickly turns to dust. Dust and dried-up soil are whipped up from dry areas in

GREEN WALL OF CHINA: Specially planted shrubs will form part of a 700 km live barrier, intended to hold back inner Mongolia's desertification and prevent dust storms disrupting the 2008 Olympics in nearby Beijing.

the north of the country and move in clouds that can travel enormous distances – some have reached as far as Vladivostok in Russia. In South Korea massive dust storms blowing over the border from China were so bad in 2002 that primary schools had to be closed. The storms, which have already cost China more than US $2 billion a year in lost land and productivity, threaten the livelihood of at least 170 million people, and are increasing in frequency. There are even fears that they could blight the 2008 Beijing Olympics. More than 30 per cent of the total land area of China is now being affected by desertification.

Indeed a third of the Earth's land surface is vulnerable to desertification. Over-grazing and damaging irrigation schemes have already taken their toll, but climate change may be the final straw for such areas. According to the United Nations, desertification could drive 50 million people from their homes in the next ten years.

is California's fruit bowl drying out?

It's not only the poorest farmers who are losing out from climate change. California's agriculture industry is worth an estimated US $30 billion and employs more than 1 million workers. The most populous state in the United States, California has a varied climate, from snowy mountains to desert heat, and the region is no stranger to extremes of weather. But global warming is likely to make those extremes much worse.

Fruit-growers know how sensitive trees are to changes in temperature. The trees need a cold period to allow buds to set, and then a steady warm period for fruit to grow. In hot weather fruit matures more quickly, producing earlier harvests and smaller fruit. Californian grapes are vulnerable to high temperatures, which can lead to premature ripening that affects the quality of the wine. The problems facing US farmers are recognised by Government. The US Congress in 2007 approved a US $3 billion agricultural disasters relief fund, specifically for farmers affected by weather events.

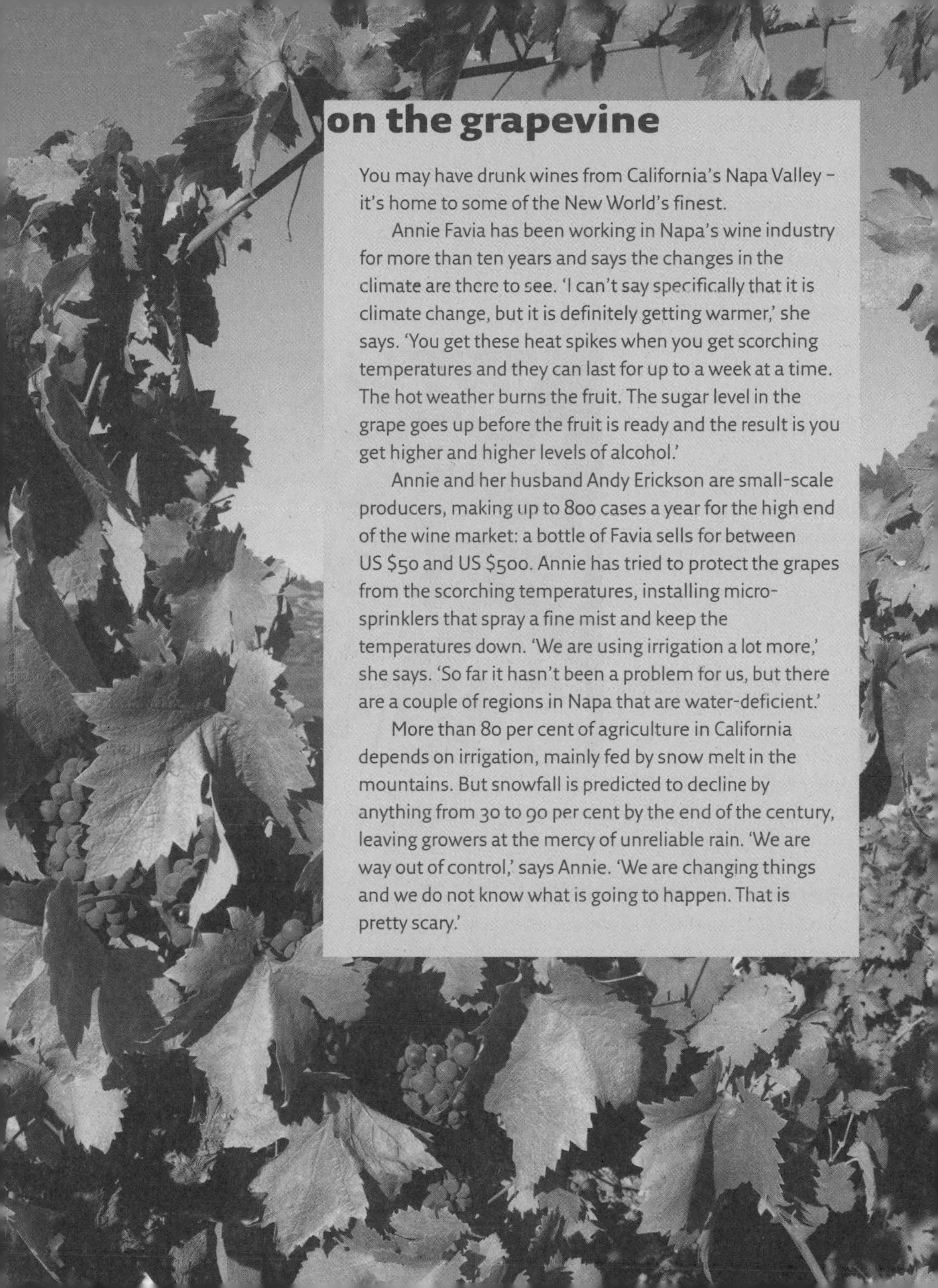

on the grapevine

You may have drunk wines from California's Napa Valley – it's home to some of the New World's finest.

Annie Favia has been working in Napa's wine industry for more than ten years and says the changes in the climate are there to see. 'I can't say specifically that it is climate change, but it is definitely getting warmer,' she says. 'You get these heat spikes when you get scorching temperatures and they can last for up to a week at a time. The hot weather burns the fruit. The sugar level in the grape goes up before the fruit is ready and the result is you get higher and higher levels of alcohol.'

Annie and her husband Andy Erickson are small-scale producers, making up to 800 cases a year for the high end of the wine market: a bottle of Favia sells for between US $50 and US $500. Annie has tried to protect the grapes from the scorching temperatures, installing micro-sprinklers that spray a fine mist and keep the temperatures down. 'We are using irrigation a lot more,' she says. 'So far it hasn't been a problem for us, but there are a couple of regions in Napa that are water-deficient.'

More than 80 per cent of agriculture in California depends on irrigation, mainly fed by snow melt in the mountains. But snowfall is predicted to decline by anything from 30 to 90 per cent by the end of the century, leaving growers at the mercy of unreliable rain. 'We are way out of control,' says Annie. 'We are changing things and we do not know what is going to happen. That is pretty scary.'

climate crisis in Africa

Farming is the backbone of most African economies. Four out of five Africans live in the countryside and farm or keep livestock for their livelihoods. As extreme weather hits the continent some of the world's poorest people are bearing the brunt of climatic shifts. Eastern and Southern Africa have been badly affected by changes in rainfall. Until recently the rainy season would normally arrive in April and May but it hasn't done so since 2002. Rivers and irrigation canals are running dry.

According to development agencies such as Christian Aid many people in rural areas are living on the edge of starvation. As temperatures rise, declining crop yields could leave hundreds of millions unable to produce or purchase enough food. In 2006 nearly 4 million people in Kenya needed emergency food aid and millions of cattle perished following three consecutive years of failed rains. Lack of rain and grazing for animals is destroying the way of life for cattle-herders, fishermen and farmers, who find themselves competing and sometimes in conflict over a dwindling resource. Without food or crops to sell, people are struggling. Those with livestock or access to pasture have to defend it from those without. In northern Kenya, shepherds carry automatic rifles. Others survive only with the help of food aid. Some turn to the towns and cities in search of an alternative living, often leaving children with grandparents to work the land.

SHARE ISSUE:

The early impacts of climate change are being felt by the people least responsible for bringing it on. The IPCC predicts mounting pressure on water resources, and a halving of yields from rain-fed agriculture for some African countries by 2020.

water shortages in Kenya

'When I was young there were grazing fields, water, milk, blood and meat,' says Lore Kapisa who heads a family of 20 in Turkana, Kenya. 'But we have seen huge changes over the past ten years: our livestock have died, our grazing fields have shrunk and our water dried up.' The people in Turkana have lived in the arid terrain in the north-west corner of Kenya by farming animals. Lore has struggled for the past ten years because of relentless drought. As water and pasture become scarcer, disputes between neighbouring groups are spiralling into violence and Lore now carries an automatic rifle to protect his herd. But rather than food aid Lore wants water, pasture and a vet. This, he says, would enable them to continue their traditional ways of life. 'We have been sick and without food, but we are human beings capable of being productive. Food aid creates dependency and reduces us to lesser human beings.'

the science of climate change

chapter 2

We know climate change is happening, but what's causing it? This chapter looks at the facts behind the freaky weather, and the implications for our planet if things carry on as they are.

why climate change is happening

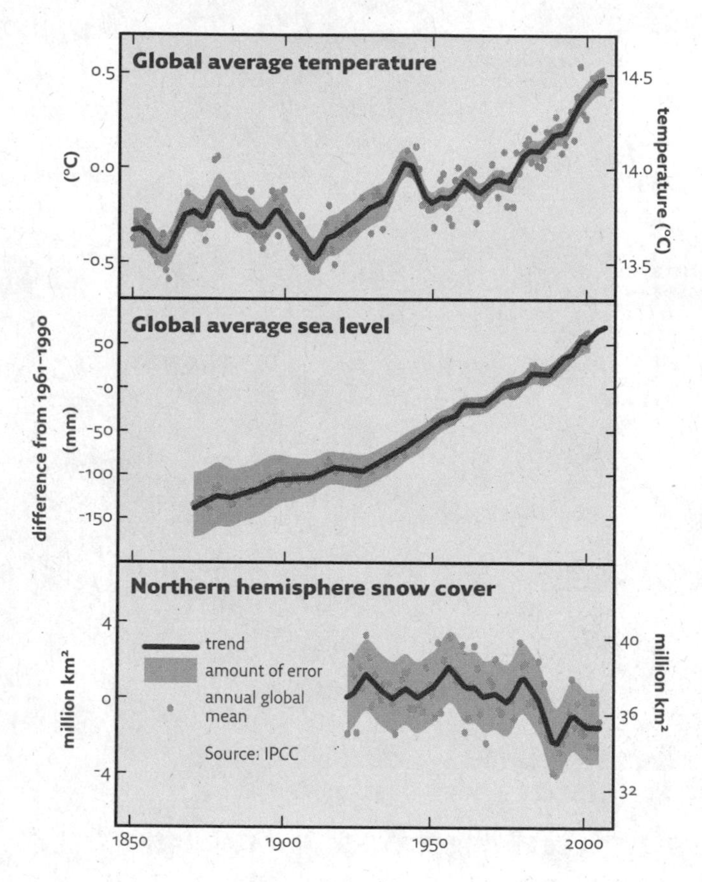

Vital signs

Three charts that tell a story – as global temperatures and sea levels have risen snow cover in the northern hemisphere has shrunk. The red lines give a bird's eye view of key changes over the past 150 years based on yearly measurements (the grey dots), while the narrowing blue bands shows the room for doubt declining as measurement techniques have improved.
Source: IPCC.

the greenhouse effect

Most life on Earth relies on energy from the sun to provide warmth and light. Some of the sun's energy reflects off the Earth's surface in the form of infrared rays, and gases in the atmosphere trap and/or reflect this energy - just as glass keeps heat in a greenhouse. The greenhouse effect helps keep the Earth warm enough to support life.

But some of these gases, particularly carbon dioxide, are building up because of pollution created by human activity. The result is more trapped heat and a sharp rise in the rate of warming.

SUN TRAP:

A thin blanket of gases around the Earth prevents some of the sun's energy from escaping back into space.

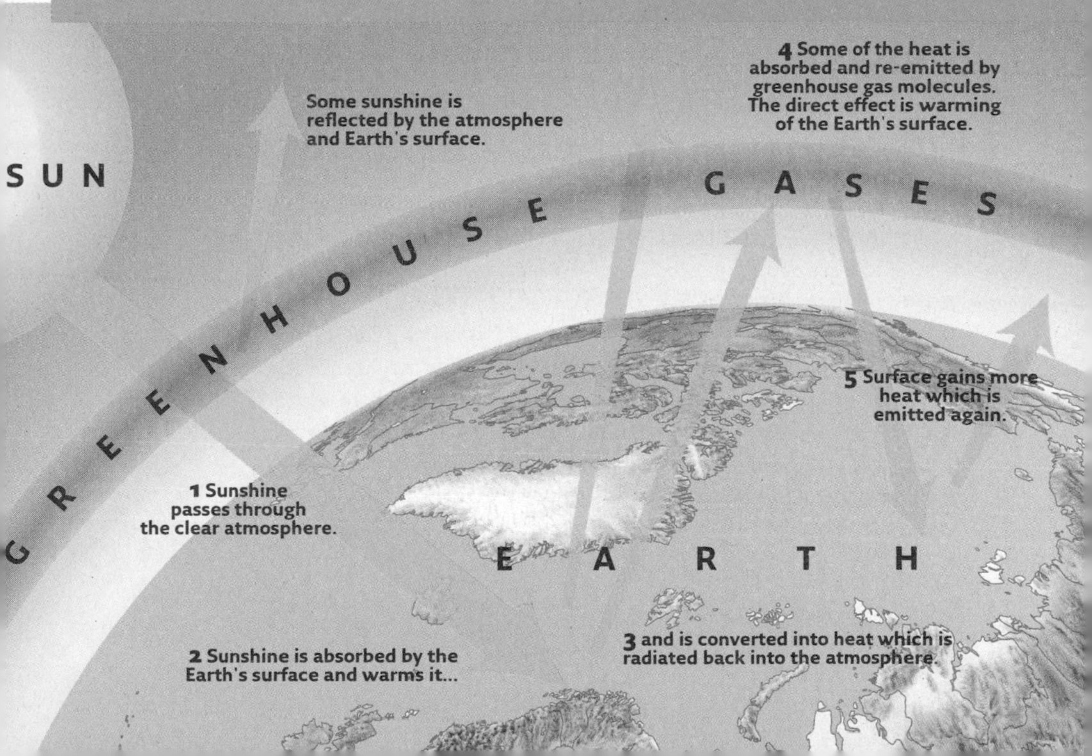

SUN

Some sunshine is reflected by the atmosphere and Earth's surface.

4 Some of the heat is absorbed and re-emitted by greenhouse gas molecules. The direct effect is warming of the Earth's surface.

GREENHOUSE GASES

5 Surface gains more heat which is emitted again.

1 Sunshine passes through the clear atmosphere.

EARTH

2 Sunshine is absorbed by the Earth's surface and warms it...

3 and is converted into heat which is radiated back into the atmosphere.

five warming signs

Higher temperatures: Scientists have established that the global average temperature has increased by 0.76°C in the past century. Records going back over 150 years show that globally 19 of the 20 hottest years have occurred since 1980.

Melting ice: Arctic and Antarctic ice is thinning and sections of ice shelves are breaking off completely. Temperatures in the Arctic are rising twice as quickly as the global average.

Coral bleaching: Scientists have found a rise of 2°C can kill coral. Reefs are home to around a quarter of known marine species.

Rising seas: We saw an increase of 17 cm during the 20th century.

Drier: Droughts have become more intense over the past 30 years, and have lasted longer, particularly in the tropics and sub-tropics.

fossil fuels and the greenhouse effect

Plants, trees and ocean plankton containing carbon absorbed many millions of years ago fossilised underground to form oil, coal and gas. When these are burnt the carbon is released, and combines with oxygen to form carbon dioxide. The first person on record to recognise the power of fossil fuels to change the climate was Svante Arrhenius, a Swedish scientist who published his 'greenhouse law' in 1896. Arrhenius estimated that doubling carbon dioxide levels in the atmosphere would lead to a rise in temperature of 5 °C and thought this promised a warmer climate in colder parts of the world. His maths wasn't far off but few people today would agree with his predictions that rising temperatures are a good thing.

greenhouse gases

The greenhouse gases include carbon dioxide, methane, nitrous oxide ozone, water vapour, sulphur hexafluoride and halocarbons. They make up quite a small proportion of our atmosphere; some have a more powerful warming effect than others.

carbon dioxide

Carbon dioxide (CO_2) is naturally occurring. People and animals produce it when breathing out whereas plants absorb and metabolise it, releasing oxygen in return.

The level of carbon dioxide in the atmosphere is measured in parts per million (ppm); or simply by weight (kg, tonnes).

Scientists believe carbon dioxide levels have been much higher in the Earth's distant past. Geological evidence suggests

RING LEADERS:

Trees are in the frontline of defence against climate change – the vast sub-arctic forests of the north being among the most important. The United States, for instance, would be having an even greater impact on global warming if its forests weren't absorbing around a tenth of its carbon dioxide emissions.

that in the Cretaceous period (65–144 million years ago) it was three to six times higher than today. The world was also 10–15 °C warmer then and there was no ice at the Poles. For much of the time that people have been on Earth, carbon dioxide levels have been stable at around 270–280 ppm. But since the Industrial Revolution they have been rising quickly, reaching 380 ppm in 2005. This is largely a result of our use of coal, oil and gas: producing, distributing and burning fossil fuels accounts for three quarters of all the emissions of carbon dioxide caused by humans. The rest comes from changes in the way land is used, particularly cutting down and burning trees. Scientists have calculated that carbon dioxide is responsible for 63 per cent of global warming. And it can stay in the atmosphere for up to 200 years. Its long lifespan means that carbon dioxide released today will still be affecting the climate for hundreds of years.

Carbon moves between land, sea and air. This cycle was in a finely tuned natural equilibrium, but it's being disturbed by humankind's use of fossil fuels, land use change and deforestation. This diagram shows where and how much carbon is stored and the yearly changes in billions of tons of carbon (GtC).

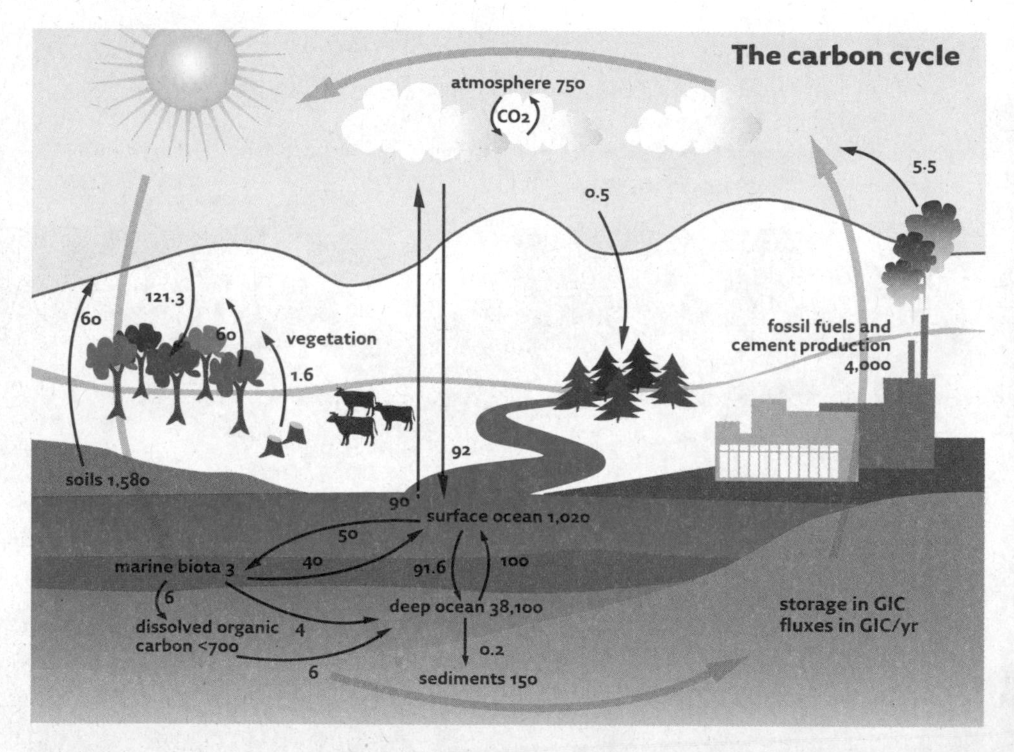

CHILDREN OF THE REVOLUTION:

With their economies run on fossil fuels, the countries of the developed world have been responsible for 77 per cent of carbon dioxide emissions since the start of the Industrial Revolution.

carbon and carbon stores

Often described as the building block of life, carbon is present in all living things. It is the fourth most common element in the universe and is found in millions of different compounds. Commonly found in mineral form as coal, it is also in oil and methane (the main constituent of natural gas). When burnt, carbon combines with oxygen to form carbon monoxide (CO) and, more often, carbon dioxide (CO_2). The world's oceans, forests and soils absorb huge quantities of carbon, in the form of carbon dioxide, from the atmosphere.

Plants take carbon dioxide out of the atmosphere through photosynthesis. When a tree is cut down and burnt, some of the stored carbon in the tree converts back into carbon dioxide and escapes into the atmosphere.

Oceans absorb around half of the carbon dioxide in the atmosphere mostly as dissolved bicarbonate . Like plants, plankton in the sea take up carbon dioxide through photosynthesis. But as oceans warm they absorb less. The increasing concentration of carbon dioxide in the atmosphere is also increasing the acidity of the oceans and damaging marine life.

Organic material in soil takes carbon dioxide out of the atmosphere. More carbon is held in the world's soils than in the atmosphere. But the capacity of the soil is limited and over time the amount of carbon dioxide escaping into the atmosphere will increase.

methane

Methane (CH_4) is a naturally occurring greenhouse gas but levels in the atmosphere have more than doubled since the pre-industrial era. Farming contributes a huge amount: one dairy cow produces an estimated 500 litres of methane daily, mainly when burping – and there are some 10 million cattle in the UK. But methane also comes from landfill sites, burning fossil fuels, wetlands and drying peat bogs (swamp gas). It has a relatively short life span, of 11–12 years. Even so, methane is a powerful greenhouse gas, absorbing 20–25 times more infrared energy than carbon dioxide, and it is responsible for 24 per cent of global warming.

AN ILL WIND:

Domestic animals produce about a quarter of the world's methane emissions – around 100 million tonnes annually. That's more in carbon dioxide equivalent than emissions from transportation.

nitrous oxide

Like other greenhouse gases levels of nitrous oxide (N_2O) have increased as a result of human activity. Chemical fertilisers and the burning of wood and fossil fuels are contributors. Nitrous oxide stays in the atmosphere for around 150 years and is responsible for about a tenth of global warming.

halocarbons

CFCs (chlorofluorocarbons), HCFCs (hydrochlorofluorocarbons) and HFCs (hydrofluorocarbons) – a group of greenhouse gases collectively known as halocarbons – have been building up in the atmosphere as a result of industrial processes. For many years CFCs were used in aerosol sprays and fridges. But scientists realised that these gases were damaging the Earth's ozone layer – which blocks out harmful ultraviolet radiation from the sun. Public opinion and environmental campaigners helped persuade politicians to phase out CFCs. Although some of the gases used to replace them (HFCs and HCFCs) are also greenhouse gases, they are responsible for less than 3 per cent of global warming.

Carbon dioxide is the main greenhouse gas

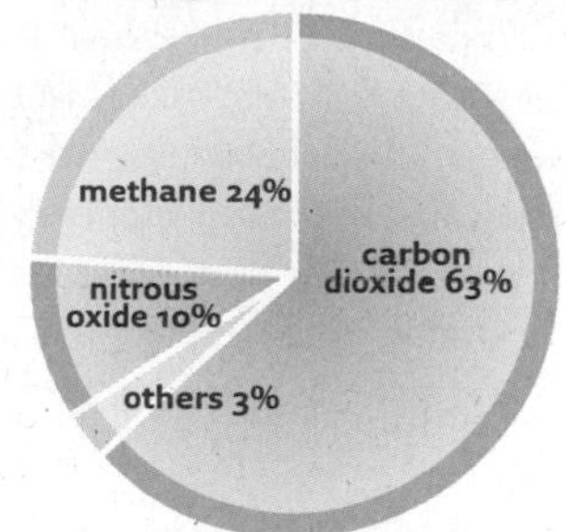

Each greenhouse gas has a different capacity to cause global warming. This, combined with the actual amounts in the atmosphere, gives the 'global warming potential' – shown in this case for the next 100 years. Carbon dioxide is the main culprit, followed by methane.
Source: Met Office Hadley Centre

the montreal protocol

The Montreal Protocol on Substances that Deplete the Ozone Layer has been described as the most effective international agreement designed to tackle environmental pollution, with developing countries being offered money to help phase out CFCs. Twenty years after it was signed CFC levels began to fall and experts are now predicting that the ozone layer will recover in the next 100 years.

ozone

Ozone in the upper atmosphere acts as a greenhouse gas while in the lower atmosphere it is part of summer smog. It may have an indirect effect on climate change. High concentrations of low-atmosphere ozone prevent plants taking up carbon dioxide.

water vapour

Although water vapour is a greenhouse gas – and the most abundant of them – human activity has only a small direct effect on the amount of it in the atmosphere. The precise role of water vapour and clouds in global warming is unclear, but scientists know that as the atmosphere warms it can hold more water vapour, causing more clouds to form. Clouds in turn can have an insulating effect, trapping warm air. But clouds may also cool things down by reflecting sunlight away from the Earth.

BLUE SKY SINKING:

The oceans absorb carbon dioxide from the atmosphere in a cycle that takes around 1,000 years to complete. The extra pollution thrown into the atmosphere by humans is distorting this process.

rising levels of greenhouse gases

Levels of greenhouse gases in the atmosphere are now more than a third higher than they were in pre-industrial times.

Scientists have compared recent rises in temperature and levels of greenhouse gases with historic data based on ice samples and tree rings. On a graph this data produces a 'hockey stick'-shaped curve depicting a rapid increase in temperature from 1900 onwards. Levels of carbon dioxide are now rising faster than at any time in the past 20,000 years.

In 2007 the Intergovernmental Panel on Climate Change (IPCC) concluded that human activities are very likely to have been responsible for most of the global warming in the past 50 years. And because some greenhouse gases stay in the atmosphere for many years, current emissions will go on warming the Earth for centuries.

Rising levels of greenhouse gases – 'the hockey stick' curves

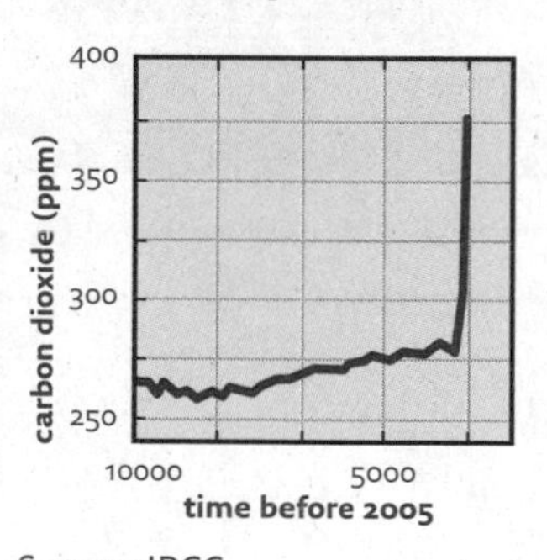

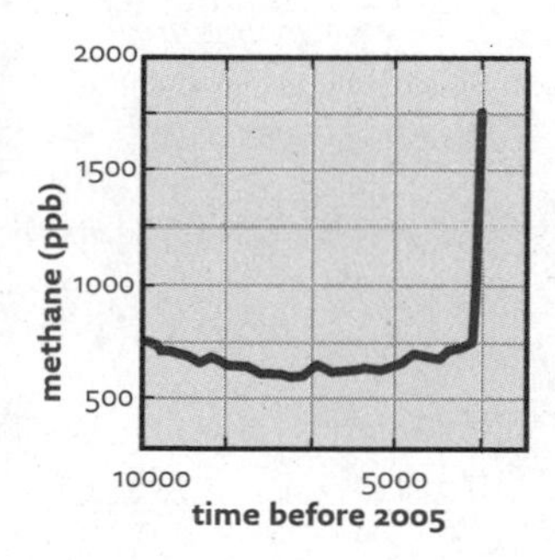

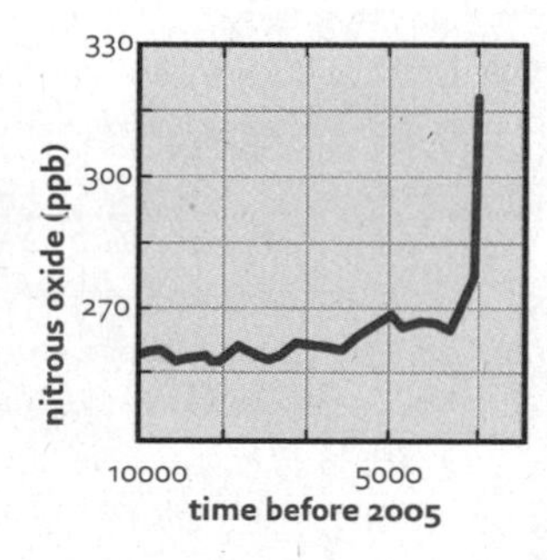

Source: IPCC

keeling curve

In 1958 US scientist Dr Charles D Keeling at the Mauna Loa Observatory in Hawaii began measuring the concentration of carbon dioxide in the atmosphere. Plotted on a graph, his results form a rising curve now known as the Keeling Curve. This shows that carbon dioxide levels vary with the seasons – by as much as 3 per cent over 12 months – because many trees in the Northern Hemisphere do not take up carbon dioxide during winter. Over and above this variation, however, we can detect a steady increase in the past 50 years.

Dr Keeling was the first scientist to report that carbon dioxide in the atmosphere was increasing, alerting the world to the threat of climate change.

dimming – hot or cold?

Whereas greenhouse gases warm the atmosphere, aerosols and solid particles reflect heat and light from the sun and prevent sunlight reaching the ground. This has a cooling effect – sometimes referred to as global dimming – and is caused by pollution from industrial processes, transport, wood-burning stoves, forest fires and volcanoes. Conversely, particles such as dust and soot can also increase warming by absorbing sunlight.

No one knows the precise impact of such pollution, but it may be masking the full impact of the greenhouse gases. It may also play a role in cloud formation, causing further complex influences on the climate. Areas downwind of pollution sources may suffer regional impacts. A recent study observed that a brown cloud of pollution over the Indian Ocean was warming the sea and could be contributing to melting of glaciers in the Himalayas.

snapshot of our future climate:

- **less snow and ice**
- **more severe storms and floods, particularly along coasts**
- **more rainfall at higher latitudes**
- **less rainfall over land in the tropics – more drought**
- **less predictable winds, rain and temperature**
- **more heatwaves**

the threat of extreme change

Some scientists believe that warming presents a risk of more extreme irreversible climate change as the Earth's natural systems swing out of balance.

Thawing permafrost: Peat bogs in Siberia and Alaska are thawing, releasing carbon dioxide and methane that they have stored for thousands of years.

The albedo effect: melting snow and ice exposes rocks, trees and tundra which are less reflective so they absorb more of the sun's energy, adding to warming and melting more snow.

Ocean warming: Methane deposits, known as clathrates, are stored in sediments under the oceans. Warming might lead to the release of huge quantities of methane gas into the atmosphere.

Scientists say there is a greater than 90 per cent chance that the great ocean conveyor – the Meridional Overturning Circulation – will slow down this century, and it could switch off in the longer term. This would imply a huge change in natural cycles.

Recent studies by NASA director and scientist James Hansen suggest that ice sheets can melt much faster than the thousands of years envisaged by the IPCC. This is because melt water under the ice sheet speeds up the breaking up and melting of the ice. Hansen also found that other greenhouse gases – and soot particles – are speeding up melting in the Arctic. This is because the dark flecks absorb more heat. As a result, he warned, an average temperature rise of just 1°C could lead to a sea level rise of between 2 and 6 metres. If Greenland's ice sheets melt quickly this will change the temperature in the Atlantic ocean, with dramatic effects on the world's weather systems.

If a small amount of human-induced global warming triggers massive natural warming, the result is likely to be climate change that happens over decades rather than centuries.

The great ocean conveyor belt

Ocean currents redistribute heat around the planet, (below), for example bringing warm water from the tropics to the UK via the Gulf Stream, and cold water to Newfoundland, with striking effects on both countries' climates. Source: IPCC.

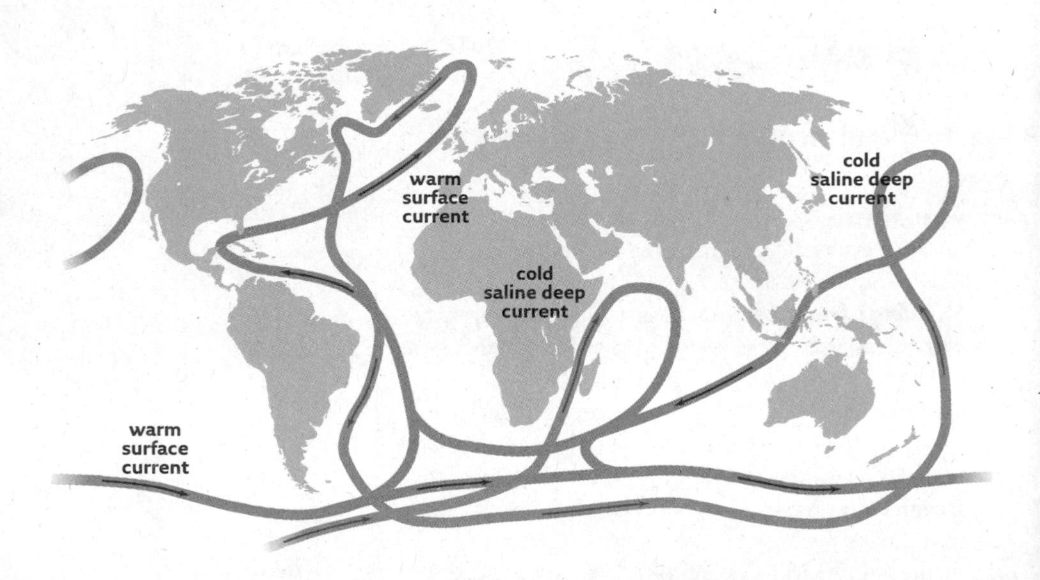

uncertainty and the sceptics

The complexity of the Earth's climate means that there has been some scientific uncertainty about global warming. Although the theory of the greenhouse effect was first put forward more than 100 years ago, clear evidence that it was happening was not readily available until the 1970s.

Today there is little room for doubt that climate change is happening and that it is caused by human activity. The IPCC – a naturally cautious body – stated in 2007, 'Most of the observed increase in global average temperatures since the mid-20th century is very likely (>90 per cent) due to the observed increase in anthropogenic greenhouse gas concentrations.'

But earlier scientific uncertainty has proved a fruitful area for media debate, amplifying the doubts of people who have questioned climate change. Some have exploited scientific doubt to spread complacency and confusion because they have seen addressing climate change as a threat to their interests. Certain companies, for example, have tried to divert attention from the environmental impact of fossil fuels.

> **" Some will always make a case for doubt in an issue such as this, partly because its implications are so frightening. But what is not in doubt is that the scientific evidence of global warming caused by greenhouse gas emissions is now overwhelming.**
>
> *Tony Blair, UK Prime Minister (1997-2007)*

Scientists have explored alternative theories on the rising temperatures – for example, that they are down to variation in the sun's cycles. But this does not explain the fit between rising levels of greenhouse gases and rising temperatures. In fact, computer models looking at the effect of the sun's output on our current climate show that if natural variations were the cause, the Earth should now be cooling, not getting hotter.

The sun's activity has in fact been decreasing since 1985. Natural phenomena simply do not explain why temperatures have risen in the past 30 years.

An assessment of more than 900 scientific studies on climate change, published over a ten-year period, found that none of the research disputed the consensus view that human activity is responsible for global warming.

fuelling the scepticism

While some scientists have pursued legitimate lines of enquiry over the science of climate change, others have deliberately exploited doubts for commercial and political gain. In the United States PR strategists advised the oil industry on how to set up groups to stir up doubts over the science and influence public opinion – in much the same way as the tobacco industry had earlier tried to persuade the public that smoking did not damage their health.

Oil has been key to the global economy for a century, and action to tackle emissions from fossil fuels has been seen to threaten the industry. So it is perhaps not surprising that politicians and many others were at first persuaded not to take global warming seriously.

the scientific consensus

The Intergovernmental Panel on Climate Change (IPCC) is an international body set up by the United Nations and the World Meteorological Organisation (WMO) to assess scientific information on climate change. It brings together climate scientists and government experts to consider research from around the world. In 2007 it shared the Nobel Peace Prize with Al Gore, the maker of *An Inconvenient Truth*.

four climate hotspots

Scientists have identified a number of hotspots around the world that could trigger fundamental changes in the global climate, including:

Amazon rainforest: Changes in the Sahara could reduce the fertility of the Amazon region, speeding up destruction of the rainforest.

North Atlantic ocean current: Melt-water from Arctic ice sheets could slow the North Atlantic ocean current, leading to cooling in northern Europe.

Asian monsoon: Changes to weather systems in the Atlantic could have a serious effect on the reliability of Asia's annual rainfall.

Sahara desert: Dust from the Sahara fertilises the Amazon, but if the region gets wetter, there will be less soil erosion and more chance of plants returning to the Sahara.

Scientists are working to understand how such hotspots could affect the climate – and what impact climate change will have on the hotspots themselves.

The IPCC's fourth report, issued in 2007, highlighted the growing evidence of observed climate change from around the world and looked at future predictions.

Its evidence is based on a consensus among the scientists involved. It is mainstream thinking – endorsed by international governments – and as such it can err on the side of caution. Some have suggested that the IPCC is in fact presenting a 'best-case scenario', underplaying the evidence for more violent climate change. One author based at the Met Office Hadley Centre Richard Betts says that means that the result is 'bullet proof' in terms of the certainty of the science it contains. 'When I read this [the IPCC report] for the first time I did feel fear – I had worked on it for three years and I knew it was right,' he says.

what will the world be like with a changed climate?

Scientists can predict with increasing accuracy how the climate will respond to rising levels of pollution in the atmosphere. In 2007 the IPCC highlighted the scientific predictions about the impacts of climate change over the next 50 to 100 years, revealing a world in which billions of people will be at risk. Sub-Saharan Africa, large river delta areas in parts of Asia, small island states and the Arctic regions are likely to be particularly vulnerable.

" It's the poorest of the poor in the world, and this includes poor people even in prosperous societies, who are going to be the worst hit and who are the most vulnerable as far as the impacts of climate change are concerned.

Rajendra Pachauri, Chair of the IPCC

the Hadley Centre's global climate models

The UK's Met Office Hadley Centre is spearheading international research. Early climate models were simplistic replicas of the climate, covering temperature, rain and carbon dioxide, and were more useful for monitoring changes in the weather than for predicting future climate. Today's models include a vast number of variables and have been shown to be capable of describing historical changes in the climate with reasonable accuracy. Interactive cloud systems, oceans, land-surface cover and aerosol pollution are all factored into the predictions made by models such as the Hadley Centre's latest (global climate model) HadGEM1. This looks at 38 different layers of the atmosphere – allowing for greater accuracy in the predictions, and more regional detail than in older models.

Africa: Many countries are already becoming drier and this is predicted to get worse. Some 75-250 million people likely to face water shortages by 2020. Farm yields in some parts will be reduced by half by 2020. Greater risk of flooding near the Equator. Sea-level rise will affect some coastal areas by the end of the century. A quarter of Africa's population lives in coastal areas.

Continental temperature trends

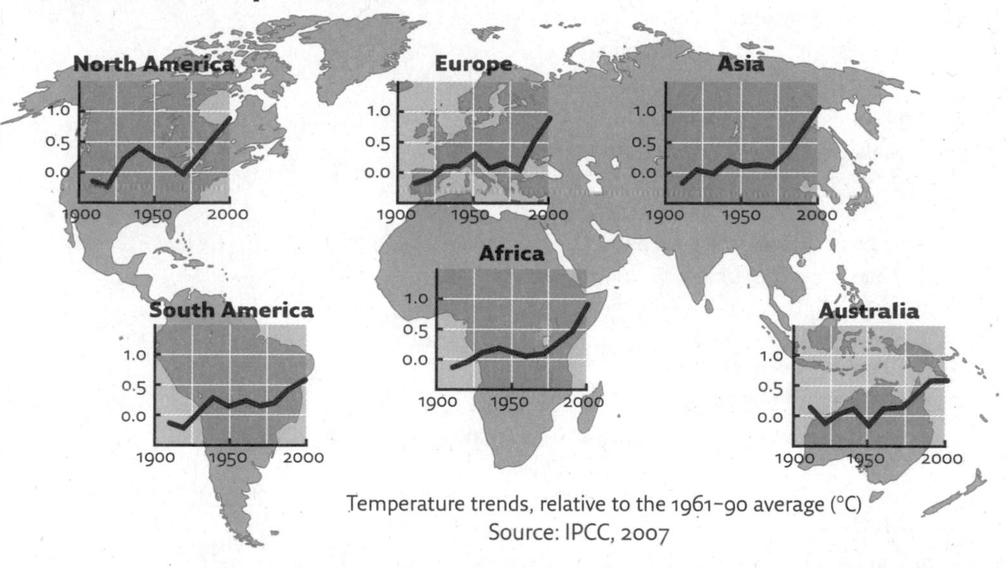

Temperature trends, relative to the 1961–90 average (°C)
Source: IPCC, 2007

Asia: More than 1 billion people facing reduced supplies of fresh water in large river basins by mid-century. Deltas in China, India and Bangladesh at risk from rising seas. Salt water likely to contaminate the ground water. Floods may bring diseases like diarrhoea and cholera. Warming oceans will affect fish and fishing communities.

Europe: More flash floods and greater coastal erosion. In Southern Europe heat and less rainfall will be bad news for farming. Human health at risk from heat waves and smoke from wildfires. Drier summers in Central and Eastern Europe will put water supplies under stress. Forests and peat areas at risk from wildfires. In Northern Europe growing seasons likely to increase, and forest likely to flourish. Less need to heat homes but risk of more winter floods.

South America: Tropical rainforest in Eastern Amazonia likely to be dry grassland by mid-century. Species such as black spider monkeys, bearded sakis, red-handed tamarins, jaguar and pumas, will come under severe threat and some truly magnificent and significant species are likely to be lost for ever. Drier, and in some areas, saltier conditions could reduce food yields. Warmer conditions could boost soybean production in temperate areas. Vanishing glaciers will threaten water supplies in cities such as Lima, Peru, as well as hydroelectric power and farming.

North America: Farmers in northern regions could have longer growing seasons for a few decades. California may dry out and become more vulnerable to winter floods. Hotter weather and wildfires will pose health hazards. Parts of the eastern coast will see more intense hurricanes and tropical storms, storm surges and flooding.

Australia and New Zealand: Australia sees increasing water shortages by 2030 and significant loss of animal and plant varieties by 2020, particularly from the Great Barrier Reef and the Queensland Wet Tropics. Water shortages predicted for New Zealand's North Island and rising temperatures will benefit farm yields for a time on the South Island. Risk of coastal flooding from mid-century in Australia and New Zealand.

climate change in the UK

With UK temperatures expected to increase by 2–3.5 °C by the 2080s, hot summers will be more frequent and very cold winters rare. Winter flooding is predicted to become more common, while farmers face shifts in growing seasons.

Sea levels will go on rising by 26–86 cm in the South East by 2080. Extreme high-water levels are likely to become more frequent, particularly on the east coast of England, and houses on floodplains are especially vulnerable. The Thames Barrier is likely to need replacing by 2030 if today's levels of protection are to be maintained. Flooding is likely to hit the least well-off most: the most deprived 10 per cent of the population are eight times more likely to be living in the coastal floodplain than the wealthiest 10 per cent.

A study of UK wildlife found that eight species are under threat from climate change because of changes to habitat. Increased drought could mean fewer slugs and snails available as food for song thrushes in the south of England, with the population at risk across England, Wales and Ireland if warming continues. Skylarks, black grouse, common scoter and capercaillie are also likely to see their preferred habitat disappear. Stag beetles, currently only found in southern England, could find more suitable habitat further north. A warmer South East will attract new species: mosquitoes are already on the up and there are fears that malaria carriers could soon be in the UK. Some species will find they cannot adapt: the Snowdon lily could die out while the snow bunting is unlikely to be found in the UK.

UK climate change in the 21st century

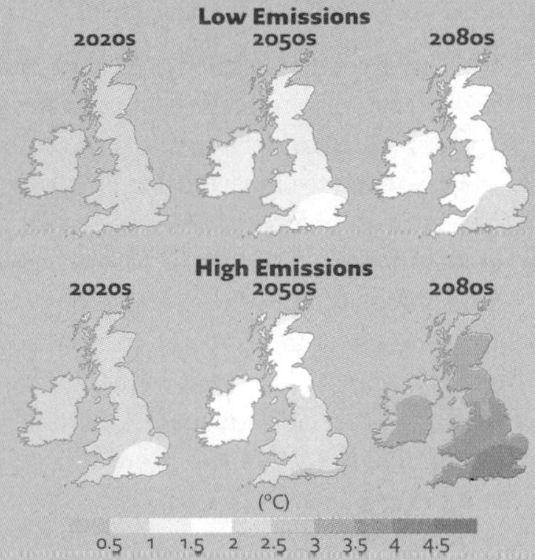

Projected temperature increases, based on one of two IPCC scenarios – Low Emissions (+2.0°C, 525 ppm CO_2 in 2080s) or High Emissions (+3.9°C, 810 ppm CO_2 in 2080s)

the human cost

The human costs of inaction on climate change are incalculable. Millions of the world's poorest people will experience dramatic changes in their way of life.

Regions close to the Equator – many of which rely on small-scale farming for food – will see bad harvests from a temperature rise of just 1–2°C. Farmers may be able to adapt by using different crop varieties and relying more on irrigation. But studies suggest that temperature increases above 3°C will be hard to accommodate. Heat will be bad news for dairy farmers: cows are less fertile, produce less milk and do not live as long in hot conditions. Cattle and pigs are also affected by heat – they tend to breed less.

Adapting to climate change will be difficult for small and family farms, for pastoralists and people who make a subsistence living on poor land, especially in parts of Africa and Asia. Without adaptation, crop yields are likely to fall and be badly affected by extreme weather events. Wealthy areas rich in resources are likely to adapt more easily.

Fishing communities will struggle as fish species migrate or die out. River fish will be affected as rainfall and snow melt patterns change. Millions of people, particularly in some of the poorest communities, rely on fishing to supplement their families' diet.

Food will not be the only area of the economy that is damaged. The timber industry will see extreme weather and more wildfires, insects and pests.

Thousands – if not millions – of people have already left their homes because of the changing climate. The International Red Cross says 25 million people could already be classified as 'environmental refugees' in 2001 and it has been estimated that climate change could push the total number of displaced people worldwide to 1 billion by 2050. One study estimated that around 15 million were likely to be displaced from Bangladesh alone. In China 4,000 villages are likely to be abandoned as a result of the spread of the Gobi desert. Huge numbers of people on the move are thought likely to increase the risk of conflict in some areas where resources are most scarce.

health risks

The health of millions is at risk. The very young and very old will be most vulnerable. The IPCC warns of:

- more malnutrition with long-term impacts for child growth and development
- more deaths, disease and injury due to extreme weather events; people in urban areas will be most at risk – cities intensify temperatures because buildings hold heat
- an increase in diseases, including diarrhoea and cholera as a result of water contamination caused by flooding
- more problems such as chest infections and asthma because of increased low-level ozone
- changes to the way infectious diseases spread.

The warming of colder regions will lead to fewer people dying of the cold. And changes to the climate could limit the spread of malaria in some parts of Africa. But over all, the health benefits of climate change are vastly outweighed by the negative effects, particularly in poorer areas.

Mortality and climate change

Estimated deaths in 2000 attributed to climate change (compared to a 1961–1990 baseline) show that Sub-Saharan Africa and the rest of the developing world bear the health burden of climate change.

Source: World Health Organization

the cost to wildlife

Plants, birds and animals cannot easily adjust to rapid changes – either in temperatures or in their food supply. The potential toll on the wildlife-rich rainforest of the Amazon is an extreme example. Experts believe the golden toad and harlequin frog – both native to Costa Rica – have disappeared as a direct result of climate change. Animals as diverse as polar bears, tigers, penguins and pikas are at risk. Some 70 per cent of coral in the Indian Ocean has already died as a result of the heat.

The IPCC predicts that 20-30 per cent of plant and animal species are at an increased risk of being wiped off the face of the Earth if global average temperatures rise by more than 1.5-2.5°C. Indeed some scientists put climate change at the top of the list of threats to biodiversity in many regions.

JUMPING OFF POINT:

Scientist JA Pounds says Central American frog species have disappeared due to deadly infectious diseases spurred by global warming. 'Disease is the bullet killing frogs, but climate change is pulling the trigger,' he said.

In the UK long dry periods damage food supplies for migrating birds. Breeding redshanks, lapwings and snipe dropped by up to 80 per cent at five reserves in the south of England following the dry summer in 2005. And warmer seas are leading to fewer plankton off the British coast, with an impact on the sand eel population – resulting in less food for nesting birds like puffins.

As plants, insects, birds and animals migrate so the nature of our landscape will change. Traditional English woodland, populated with oak, beech and bluebells could become a thing of the past, as sycamore becomes the more dominant species and cow parsley could force snowdrops and bluebells out.

the economic cost

It's easy to see the immediate costs of clearing up damage caused by floods, storms or warping railway tracks; less obviously, farmers across the world are paying the price in lost harvests, lower yields and higher prices. People who rely on rainforests for food, farming and forestry are losing their way of life.

In the developed world, the insurance industry generally picks up the tab for extreme weather. But in Europe the cost of flood damage alone would be expected to rise by up to £82 billion a year, and the costs of Hurricane Katrina have been put at US $125 billion. Many people in the poorest countries do not have insurance policies; but even in the rich developed world, many can't afford insurance as the risk of extreme weather rises – particularly for those living in flood plains.

An economic assessment of the cost of climate change, commissioned by the UK Government and published in 2006, suggested that unchecked climate change could damage economic well-being worldwide by at least 5 per cent and by as much as 20 per cent. In other words a fifth of the world economy is at risk.

'The impacts of climate change are not evenly distributed – the poorest countries and people will suffer earliest and most.'
The Stern Review, October 2006

how different temperature rises could affect us

The IPCC predicts temperature increases of 1.8-6.4 °C this century, depending on the amount of fossil fuel used. Its best estimate is 4 °C. While rises at the lower end of this spectrum would have terrible impacts for millions of people around the world, a rise of more than 6 °C could spell shocking global consequences.

HOT TIP:

A future with extreme climate change? Don't go there. Experiments in South Africa's Succulent Karoo desert show that a 4-6 °C rise in daytime temperature kills three quarters of plant species.

the six-degree scenario

Evidence suggests that the end-Permian extinction, 251 million years ago, may have been triggered by a rise of temperatures of 6 °C: it almost wiped out life on Earth. Forests, swamps and savannahs were washed away, land turned to desert; the warming oceans, which lose oxygen as temperatures rise, would have become stagnant and toxic. Some scientists believe the warming of the ocean would have been enough to trigger the release of huge clouds of methane from the sea bed, poisoning the atmosphere. Life on Earth did survive 6 °C of warming, but those changes took place over 10,000 years. Human releases of carbon dioxide are almost certainly happening faster than any natural releases since the beginning of life on Earth.

Writer Mark Lynas has studied historical records to examine the potential effects of temperature rises of up to 6 °C. Here is a summary of his six-degree scenario:

+1 °C: Deserts spread across parts of the United States, turning farmland to dust from Canada in the north to Texas in the south. The Gulf Stream could switch off – plunging Europe into an icy winter. Coral reefs around the world are wiped out.

+ 2 °C: Oceans turn increasingly acidic, killing off plankton and affecting sea life. European summers are plagued by heat waves as strong as the killer of 2003. Wildfires spread around the Mediterranean. Greenland tips into irreversible melt, accelerating sea-level rise and threatening coastal cities. The polar bear and walrus become extinct.

+ 3 °C: Deserts spread across parts of Africa, driving millions of refugees to surrounding countries. A permanent El Niño grips the Pacific, causing weather chaos worldwide. Drought and wild fires rage across the Amazon, destroying swathes of forest, releasing yet more carbon. World food running short. Water shortages threaten parts of India, Pakistan, Australia and Peru.

+4 °C: Tens of millions become refugees as rising waters threaten the Nile Delta and low-lying Bangladesh. The West Antarctic ice sheet collapses, pumping 5 metres of water into global sea levels. Southern Europe becomes like the Sahara, with deserts spreading in Spain and Portugal.

+5 °C: The Earth is hotter than at any time for 55 million years. Desert belts expand across Europe, America and Asia. Some populations try to move towards the poles. Most of the world is uninhabitable.

+6 °C: Huge firestorms sweep the planet as methane hydrate fireballs ignite. Seas release poisonous hydrogen sulphide. Most of life on Earth has been extinguished. Humanity's survival is in question.

how much carbon can we live with?

Climate change will get worse if we allow greenhouse gas emissions to grow unchecked. So where are the emissions coming from? And how much more can the planet take?

chapter 3

scientific predictions

Models show that if we go on with business as usual and fail to address greenhouse gas emissions we can expect average temperatures across the world to rise dangerously. Precisely how much warmer it will get is the subject of huge investigation but the Intergovernmental Panel on Climate Change (IPCC) gives a range of between 2.4°C and 6.4°C above pre-industrial levels by the end of this century. If we don't take action, scientists believe the most likely point within that range will be a rise of 4°C by the end of this century – that's enough to produce catastrophic results.

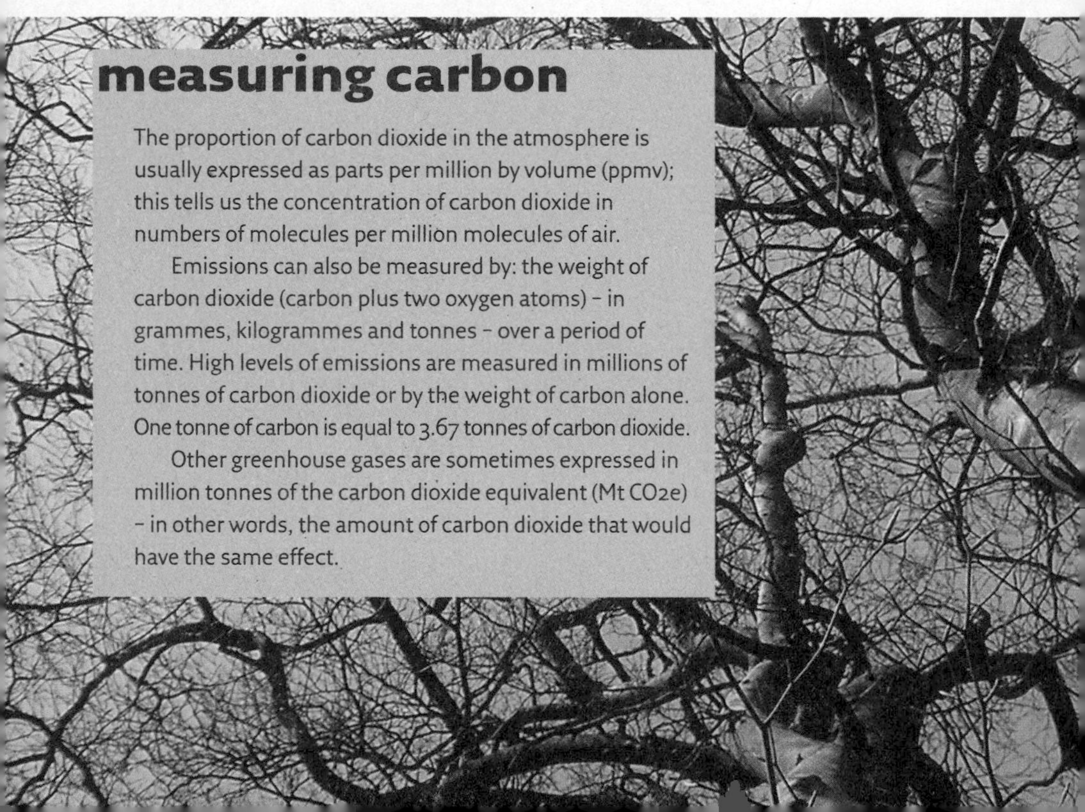

measuring carbon

The proportion of carbon dioxide in the atmosphere is usually expressed as parts per million by volume (ppmv); this tells us the concentration of carbon dioxide in numbers of molecules per million molecules of air.

Emissions can also be measured by: the weight of carbon dioxide (carbon plus two oxygen atoms) – in grammes, kilogrammes and tonnes – over a period of time. High levels of emissions are measured in millions of tonnes of carbon dioxide or by the weight of carbon alone. One tonne of carbon is equal to 3.67 tonnes of carbon dioxide.

Other greenhouse gases are sometimes expressed in million tonnes of the carbon dioxide equivalent (Mt CO_2e) – in other words, the amount of carbon dioxide that would have the same effect.

how high are global emissions?

In a little more than a century human activity has added 200 billion tonnes of carbon to the atmosphere. Today we are pumping out just over 7 billion tonnes of carbon a year – that's twenty times the weight of all the water in Lake Windermere. It's also equivalent to 26 billion tonnes of carbon dioxide. A further 1.6 billion tonnes of carbon dioxide emissions each year come from changes in the way we use land – mostly by cutting down forests that, when growing, play a vital role in soaking up carbon dioxide.

Scientists agree that we need to stabilise emissions, then reduce them. But at what level?

carbon footprint

'Carbon footprint' can be useful shorthand to describe the amount of carbon dioxide we generate – individually, in our cities, nationally, or from the products we use. The term 'footprint' is also used to describe the *impact* of emissions, by providing a measure based on the area of things like forests, oceans and grassland that are required to absorb the carbon dioxide.

Later in this chapter, we'll use this measure of impact to look at the carbon footprints of different countries. Elsewhere in the book we'll be focusing on carbon dioxide emissions, particularly in the UK, where government figures for emissions are given in weight.

the 2°c target

The UK government, together with the rest of the European Union, has pledged to stabilise greenhouse gases to prevent average global temperatures rising more than 2°C above pre-industrial levels. Allow greater temperature change, climate scientists say, and we start to play havoc with agriculture, the economy and the natural environment – committing millions of people to hardship and millions of species to extinction. Perhaps even more worrying is the prediction that a higher rise than 2°C would increase the risk of spiralling and irreversible climate change.

UK path to a safer future

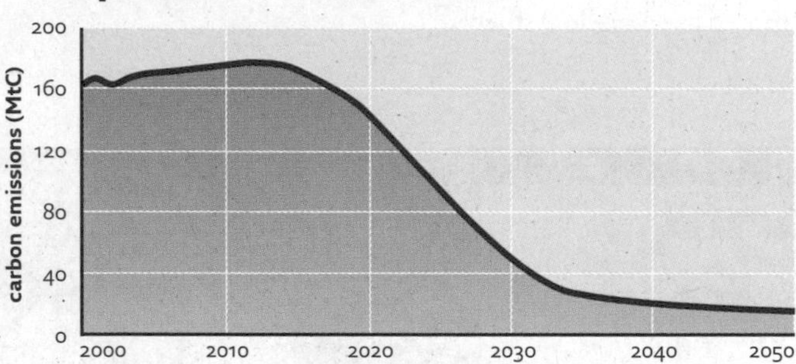

Steady but steep - the UK needs to make serious cuts in carbon dioxide emissions over the next 40 years to play its part in avoiding runaway climate change. Based on research by the Tyndall Centre for Climate Change Research at the University of Manchester, the chart assumes the need to stabilise atmospheric carbon dioxide at 450 ppm.

To stay within a 2°C rise the science tells us we would need to keep greenhouse gases (and other pollutants) at no more than 450 parts per million carbon dioxide equivalent. But we're already racing towards that threshold: carbon dioxide alone had already reached 380 ppm in 2006, and emissions are rising by nearly 2 ppm per year.

And there's more. As a result of the carbon dioxide already in the atmosphere scientists warn that 2°C could be the lowest increase we can hope for.

So how much time do we have to turn things around? Because carbon dioxide stays in the atmosphere for so long, and because it takes a while for it to affect the climate, the fossil fuel we burn in power stations today will still be affecting the climate at the start of the next century. That means we have to reduce emissions as quickly as possible, rather than hoping to make big cuts five years from now. The sooner we make cuts, the better.

Two facts illustrate the urgency of the task that faces us if we want to stay within a 2°C rise:

1 Greenhouse gas emissions have to start falling by 2015. They must then continue to decline.

2 Industrialised countries need to cut emissions by about 90 per cent by the middle of this century.

So can it be done? Some argue that it is unrealistic to expect to cut carbon emissions quickly and deeply enough to stay within 2°C. But the scientific consensus is that we must.

where in the world do the emissions come from?

Most carbon emissions from human activity come from industrialised countries, whose economies rely on manufacturing and transport powered largely by fossil fuels.

The United States, home to fewer than 1 in every 20 people in the world, is responsible for about a fifth of global carbon dioxide emissions.

Europe accounts for 15 per cent of global carbon dioxide emissions, working out at around 9.3 tonnes per person per year. That's enough to fill two Olympic swimming pools for every person in Europe.

China became the biggest emitter of carbon dioxide in 2006. But China's emissions per person are less than those of the United States: in 2003 the average American emitted nearly six times as much carbon dioxide as the average person in China.

Across the world average emissions were 3.7 tonnes per person in 2003. But in some African countries, such as Mali and Burundi, they were less than 0.1 tonnes per person.

Comparative worldwide emissions (2003)

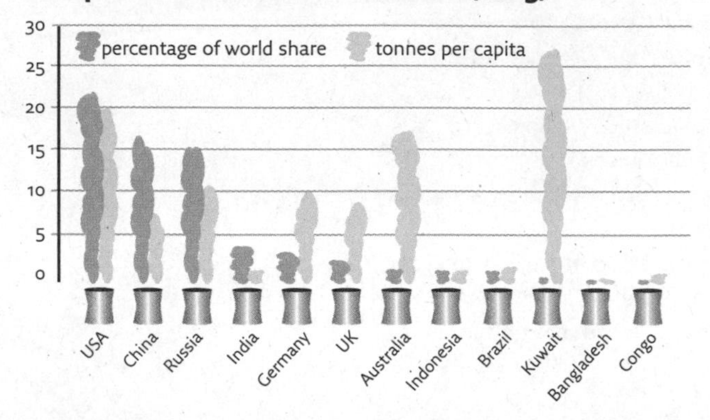

NB: Includes international aviation and shipping, excludes land use change. Source: World Resources Institute

A country's greenhouse gas emissions depend on the size of its economy, the types of industry it has and the kinds of energy it uses. Demand for heating and air conditioning are also key factors: rich countries with hot summers use lots of energy for air-conditioning; cold countries use more for heating.

China has become a major emitter of greenhouse gases only in the past 20 years while the United States and the UK have been pumping out carbon dioxide throughout the past century. And it is the total quantity of carbon dioxide in the atmosphere that is driving global warming, not just the annual amount added to this total. So it's the developed countries that are mostly responsible for the problem.

seven people who can make a difference

1. The President of the United States
2. The President of China
3. The British Prime Minister
4. The Mayor of London
5. Your boss
6. Your council leader
7. You

comparing carbon footprints

A country's real contribution to the total amount of greenhouse gases in the atmosphere depends not just on the energy it uses, but also on the way it uses land and the stuff it consumes. The amount of land being farmed will affect emissions of methane and nitrous oxide as well as the amount of carbon dioxide held by the soil. (Forests help remove carbon dioxide from the atmosphere.) Goods imported from overseas also have their own carbon footprint and use up natural resources.

To get a clear picture of a country's overall footprint, then, we need to look at what it is consuming, including imports. The Global Footprint Network uses information on all of these areas to estimate each country's total impact on the planet in terms of environmental cost, and has come up with a measure it calls global hectares. Using this measure, humanity's global footprint is put at 13.9 billion global hectares. That's 24 per cent more than the natural resources available on Earth. This is a problem: we are exceeding the planet's ability to absorb our impact.

Global footprints (global hectares/person, 2003)

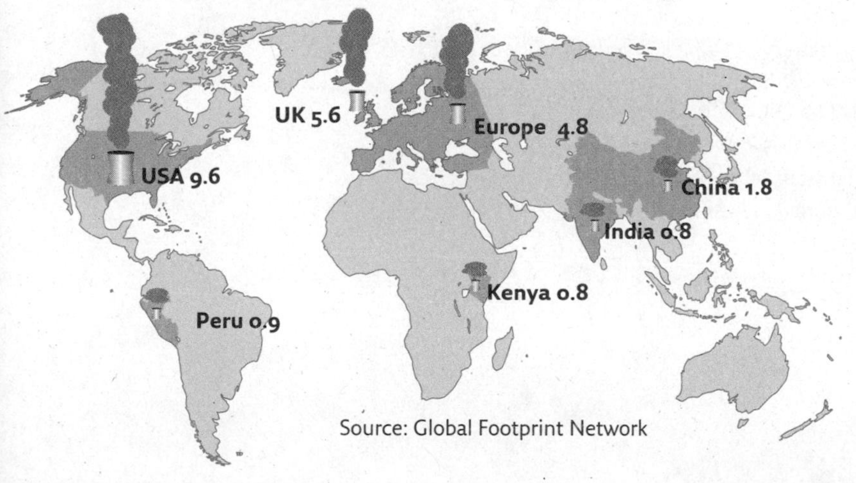

Source: Global Footprint Network

UK emissions

Half a billion tonnes. That's how much carbon dioxide the UK emits each year. Some 60.2 million people live in the UK – that's just less than one in every 100 people in the world; but in 2005 our emissions were around 2 per cent of the global total – double our fair share.

In fact the true total is even higher: 556 million tonnes is the official figure, but that doesn't include our share of emissions from international aviation and shipping; nor the emissions created by all the products we buy from overseas.

Add to that the UK's history and the footprint gets heavier still. We have been pumping out carbon dioxide since the Industrial Revolution. It is this historical record that leads many to argue that greenhouse gas emissions are a problem created by the developed countries, including the UK, and to insist that it is the old industrial world's responsibility to reduce its emissions first. Richer countries, particularly the United States, have been reluctant to cut emissions before developing nations do, claiming it will damage their economies. This has been a major barrier to international agreements to tackle climate change.

the need to cut

Politicians may argue about who is responsible for cutting just how much carbon dioxide; but to keep within the 2°C threshold, globally we need to reduce emissions of carbon dioxide massively by 2050. This looks like a tall order – but it is possible.

Where can the cuts come from? And what difference will they make to the way we live?

is it all about population growth?

Some people argue that tackling climate change means doing something about the world's growing population: the Earth's resources are limited and there simply is no longer enough to go around if we go on consuming at this rate; our survival, some say, depends on keeping our numbers down.

The global population has grown more than six-fold just over 200 years to around 6.6 billion in 2007. It is expected to stabilise at around 9.2 billion by mid-century. But the countries with the fastest growing populations are among those who have done the least to affect climate change

Carbon dioxide emissions per person are far higher in the richest countries of the world such as the United States, Canada, Japan, Australia and the UK. By contrast, many countries in Africa have per capita emissions that are negligible in comparison – less than one-hundredth the level. Many countries in Latin America have per capita emissions that are only around one-tenth those of the richest countries.

Global population is not the root of the problem: it is the rapid use of energy and natural resources by the richer and fossil-fuel dependent economies that is putting pressure on the planet.

Carbon-dating climate change – a timeline

- **End of the Permian period (251 million years ago).** Geological evidence suggests carbon dioxide concentrations were four times higher than today. Much of Central Europe and the southern United States was desert, and seas were 20 metres higher. Methane releases from the sea beds are thought to have wiped out many living things. Some estimates suggest that 95 per cent of life on Earth was eradicated.

- **Start of the Holocene (10,000 years ago).** End of the last ice age and the beginning of a relatively stable climate.

- **Medieval Warm Period (900–1200 AD).** Relative warming in the Northern Hemisphere, with wine grown in the UK and droughts in North America. Temperatures in Europe are thought to have been 1°C -2°C warmer than at the start of the 20th century, but little evidence exists of warming on a global scale.

- **1750–1800 Start of the Industrial Revolution** and the use of coal to drive industry.

- **1800 Start of the Anthropocene.** According to some climate scientists, the stable period of the Holocene is coming to an end: humanity's influence over the world's climate is so great that the current age should be known as the Anthropocene.

- **1896 Svante Arrhenius, a Swedish chemist,** identifies the global warming properties of greenhouse gases, predicting that a doubling of carbon dioxide levels could raise temperatures by 5°C.

- **1979 First World Climate Conference** highlights concerns about levels of carbon dioxide in the atmosphere. The Intergovernmental Panel on Climate Change (IPCC) (see p.54) is established.

- **1987 The Brundtland Report,** *Our Common Future*, makes the case for, and provides a definition of, sustainable development.

1990 The Intergovernmental Panel on Climate Change (IPCC) publishes its first assessment; it says climate change is a concern and human activities are likely to be contributing.

1992 Rio Earth Summit. World leaders meet in Rio de Janeiro, Brazil, to discuss the state of the planet, signing up to the United Nations Framework Convention on Climate Change.

1997 Kyoto Protocol. The first legally enforceable international treaty designed to cut greenhouse gas emissions. Signed by 141 countries, the Kyoto Protocol eventually comes into force in 2005.

2001 Newly-elected US President George Bush refuses to ratify the Kyoto Protocol. 'I oppose the Kyoto Protocol because it exempts 80 per cent of the world, including major population centres such as China and India, from compliance, and would cause serious harm to the US economy.'

2002 Larsen B ice shelf breaks up. A piece of ice a quarter the size of Northern Ireland falls into the sea in the Antarctic.

2003 European heat wave. Some 35,000 Europeans die as a result of extreme temperatures.

2004 UK Prime Minister Tony Blair is 'shocked' by scientists' warnings on climate change.
He tells reporters that 'unchecked climate change has the potential to be catastrophic in both human and economic terms'.

2005 UK Prime Minister Tony Blair makes climate change a priority for the G8 rich countries summit at Gleneagles. UK emissions continue to rise.

2006 UK Government agrees to introduce new legislation designed to tackle climate change. UK emissions continue to rise.

2007 The IPCC and former US Vice President Al Gore share the Nobel Peace prize. Gore's film *An Inconvenient Truth*, highlighting the threat posed by climate change, wins an Oscar. Japan announces that climate change will be a priority at meeting of the G8 industrialised countries.

breaking down the UK's carbon footprint

When looking at where our carbon dioxide emissions come from we can slice the cake a number of ways. The flow chart shows one breakdown by end-user. But a common way to think about emissions is in terms of three key end-uses – electricity, heat and transport.

electricity

Demand for electricity is responsible for roughly a third of the country's carbon dioxide emissions. That's because we rely on fossil fuel: large, centralised power stations burn coal and gas to generate most of our electricity. The UK does produce some electricity using renewable sources such as wind and hydropower, but at the moment this provides only around 4 per cent of our electricity.

Demand for electricity in the UK has doubled since the 1970s – partly as a result of the growing number of smaller households, most with their own fridge, freezer and microwave. Over this period power stations have become more efficient, but a diminishing supply of North Sea gas, combined with high oil and gas prices, have led generating companies to turn to coal. Coal produces higher levels of carbon dioxide emissions than gas does and can cause other pollution problems as well.

heat

If we're looking at our footprint in terms of the end use of the energy, heating homes, offices and public buildings plus the heat used in industry actually account for the biggest slice of

a snapshot of the UK's CO_2 emissions in 2005

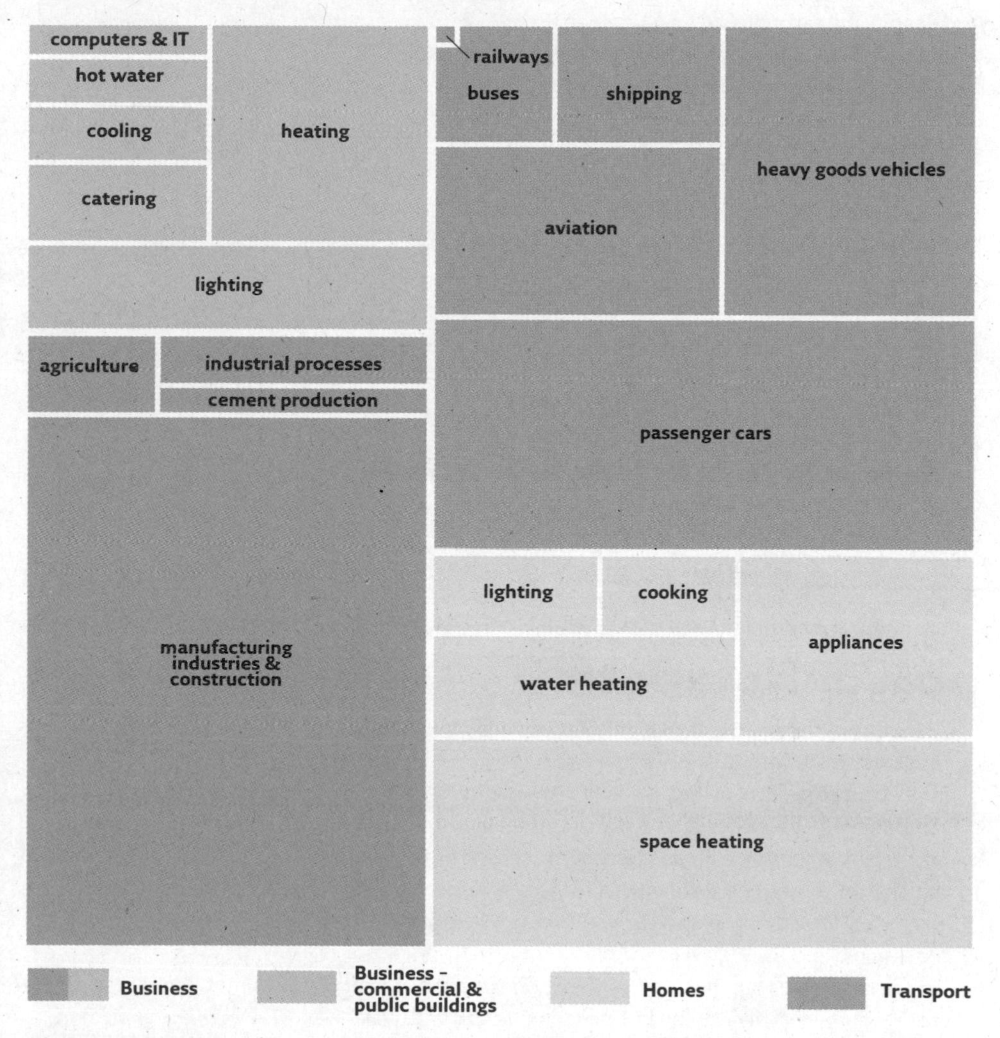

The areas of the boxes show how each sector contributes to the UK's total of 595 $MtCO_2$ in 2005. We've included international aviation and shipping. Source: BERR/Friends of the Earth.

the UK's emissions. Industry and domestic consumers alike rely mainly on oil and gas. Total demand for heat energy is falling slightly, partly thanks to improvements in efficiency. And despite the recent trend towards coal, the switch to gas that has been happening since the 1970s means that carbon emissions from heat generation have been falling. In our homes, gas is the main source of energy – and domestic use is rising slowly.

How we heat our homes

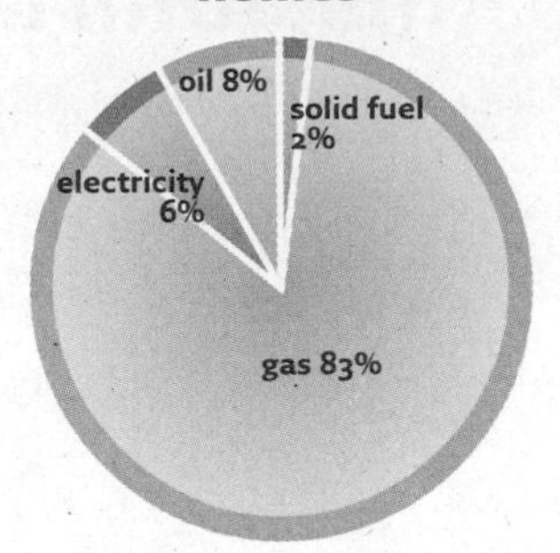

Source: BERR

Carbon emissions from industry (2002)

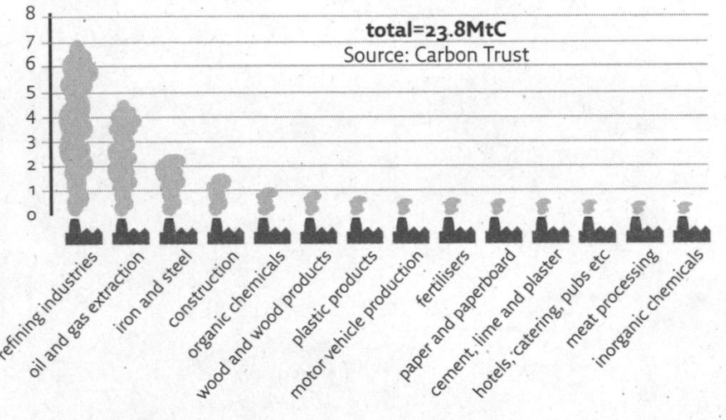

power hotspots

If we look at our emissions in terms of broad economic sectors, the biggest sources are business and transport and our homes.

About 40 per cent of the UK's carbon emissions come from businesses, with industry accounting for about half of that. The chemicals industry accounts for a fifth of all the energy used in industry, with food, drink and tobacco the second biggest industrial consumer. The UK's service economy – i.e. banking, insurance and shops – is less energy-intensive, but as the dominant sector still uses lots of heat and power. In fact services account for a fifth of our greenhouse gas emissions.

As individuals our main energy-consuming activity is heating our homes. Private transport comes second – followed by food and drink, health and hygiene and heating water. Holidays, clothing, cooking and books and newspapers come low down on the list.

transport

Transport is often referred to as the problem sector: since 1990 transport emissions have risen by a fifth. Some 22 per cent of UK greenhouse gases come from road vehicles – perhaps not surprising since running a car today is cheaper in real terms than it was ten years ago. But aviation is a fast-growing problem.

Emissions for a journey from London to Edinburgh

	kg CO_2 per passenger
1 Plane	96.4
2 Car	71
3 Rail (high speed electric)	11.9
4 Coach	9.2

According to the Department for Transport, aviation emitted 2.5 million tonnes in 2005. But this figure only includes domestic civil aviation. If the UK's share of international aviation is included (calculated as half of all flights taking off or landing in the UK), then emissions for aviation were more than ten times that. Even then a crude measure of carbon dioxide emissions from aviation does not reveal the true extent of the impact that flying has on the climate. Because aeroplanes create pollution high in the upper atmosphere, the impact of the emissions is magnified. How much of a difference this makes is still being studied, but conservative estimates suggest UK emissions from aviation account for 13 per cent of the UK's climate change impacts. Emissions from shipping are also on the rise, as more and more goods are traded around the world.

Transport emissions (2006)

Excluding international flights and shipping

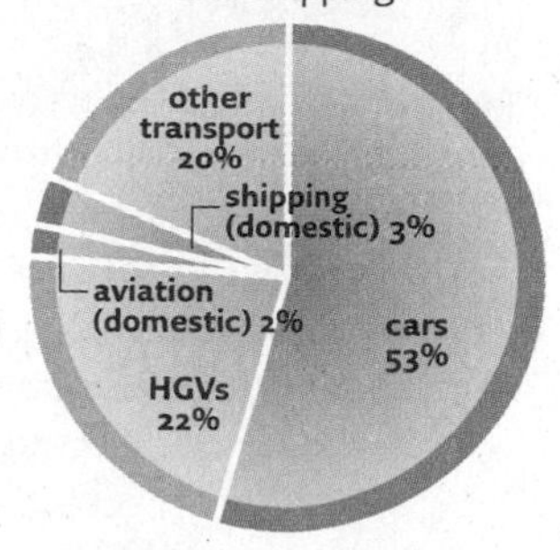

Including international flights and shipping

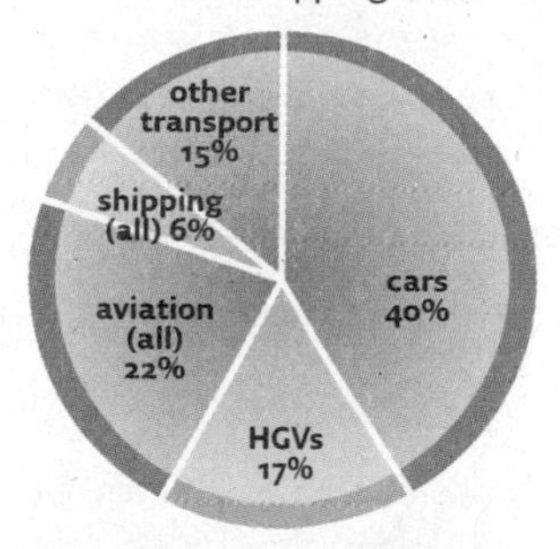

UK transport emissions are big and getting bigger. Watch out for the way they're reported: emissions from international aviation and shipping have not generally been included. But when they are it's clear how serious a problem air travel presents.

our carbon exports

None of the above takes account of the greenhouse gases the UK is responsible for elsewhere in the world. British companies play a huge role in international trade and many of the goods they sell and that UK consumers buy are manufactured overseas. Yet the energy used to produce them does not show up on the UK greenhouse gas inventory.

Take one of the UK's most successful companies, British Petroleum (BP). Its direct activities, such as exploration, drilling and refining, combined with the sale of its products, are estimated to have resulted in emissions of nearly 1.5 billion tonnes of carbon dioxide in 2004. That's more than double the carbon dioxide produced by the UK. According to one financial company, emissions from all the oil, gas and coal sold by FTSE 100 companies is likely to be close to fifteen per cent of the global total from all fossil fuels.

farmland

According to government figures, agriculture is responsible for 8 per cent of the UK's greenhouse gas emissions. This is largely from methane and nitrous oxide rather than energy use.

Intensive farming relies on chemical fertilisers, which increase the amount of the greenhouse gas nitrous oxide in the atmosphere. Fertiliser production is the largest source of nitrous oxide emissions in the world and producing fertiliser itself relies heavily on fossil fuels. Making and transporting fertilisers is the largest source of carbon dioxide emissions in the farm sector, but these emissions are not included in the government's figures for agricultural emissions. Sheep and cows emit methane as a result of the digestive process, resulting in high levels of emissions from livestock and dairy farming.

Meat farming is responsible for two thirds of the nitrous oxide and more than a third of the methane associated with human activity. One study found that every 1 kg of beef produced results in emissions of more than 36 kg of carbon dioxide equivalent: two-thirds of the energy used goes to produce and transport cattle feed. Producing food for livestock and clearing land for grazing is having a huge impact on valuable forests. Greenhouse gas emissions from livestock have been put at around 18 per cent of the global total. And demand for meat is rising, especially in the developing world: meat production is predicted to double between 2000 and 2050.

the UK's carbon footprint

Despite the government's recognition that climate change is a greater threat than terrorism, UK emission levels have continued to rise over the past decade.

Some parts of the economy have started to reduce their carbon footprint, but emissions from transport and electricity generation are rising – the latter the result of the recent switch back to coal in power stations.

The government's draft Climate Change Bill set a target of cutting carbon dioxide emissions by 60 per cent by 2050. But science suggests that this will not be enough. 'It is clear to us that climate science suggests that this figure may not be adequate to prevent global temperatures rising above dangerous levels,' said a House of Commons Joint Committee on the Draft Climate Change Bill in August 2007.

The crucial factor is not the target at some distant point in the future, but how much carbon dioxide is building up in the atmosphere. Making cuts sooner rather than later will result in less carbon in total. If the UK is committed to keeping global temperatures within 2°C it may need to look at cuts of 90 per cent by 2050.

eating our way to climate change

The contribution of food to global warming doesn't stop at the farm gate. More than 30 per cent of greenhouse gas emissions in Europe are estimated to come from the food and drink sector. In the UK the food chain contributes more than one in every five tonnes of greenhouse gas emissions.

Much of the food on our supermarket shelves has travelled the world, some of it 'air-freighted for freshness', some by ship – and almost all of it by road. A piece of beef can travel 12,000 miles just to reach the UK shops. Ninety per cent of the fruit we eat and 40 per cent of our veg are imported into the UK. Air freight has the highest emissions of carbon dioxide per tonne of food – and is being used more and more.

Even food from the UK tends to be well-travelled. The average potato's journey starts at the farm, goes by lorry to a factory or warehouse for packaging and is then taken by lorry to a centralised warehouse before distribution to the stores. One in four heavy goods lorries on UK roads are transporting food, clocking up 5.5 million food miles in 2004. That's the equivalent of one potato travelling 220 times around the Earth. But on average, lorries are only just over half full. More efficient distribution could greatly reduce the impact of food miles.

Focusing only on food miles ignores the fact that producing and processing food accounts for around 14 per cent of energy consumption by UK businesses; and food, drink and tobacco manufacturers consume more energy than is used in iron and steel production. Keeping food cool contributes to the emissions total: fresh food, especially meat, is stored and transported in refrigerated containers that produce greenhouse gases.

Globally, food production contributes around 30 per cent of total greenhouse gas emissions, half of which come from methane. And to think that on average we throw away around a third of the food we buy – where, left to rot in a landfill, it generates yet more methane.

the footprint of a packet of crisps

To discover the carbon footprint of a standard packet of Walkers Cheese & Onion Crisps, the Carbon Trust plotted out the key stages of Walkers' supply chain, from sowing the potato and sunflower seeds (for the oil) through to getting the crisps on the shelves and disposing of the packet. By looking at the energy consumption of each stage, and converting this into emissions, it worked out the total carbon footprint. Result: a packet of Walkers Cheese and Onion Crisps has a footprint of 75 g.

business as usual?

Do nothing to cut carbon dioxide emissions and the amount in the atmosphere will just build up. Climate modellers describe this scenario as 'business as usual'. It assumes that emissions continue to rise at current rates – assuming continuing growth in the economy, and the on-going use of fossil fuels. If this were to happen, carbon dioxide emissions from fossil fuel would more than double by 2050 globally. Yet if we reduce the amount of energy we use, or start to switch to non-fossil fuel energy, we will slow down the rate of growth in emissions. It may sound obvious but the faster we make changes, the quicker carbon dioxide levels will stabilise.

'The investments made in the next 10-20 years could lock in very high emissions for the next half-century, or present an opportunity to move the world onto a more sustainable path.'
Stern Review

how can the cuts be achieved?

There are many ways to curb greenhouse gas emissions. But the biggest savings will come from two simple strategies: using less energy and using cleaner fuels. We look at these more closely in the next two chapters.

chapter 4

saving energy

Saving energy is a great way to shrink our carbon footprint. In fact it's the first thing we should do. Why spend on power that we don't really need?

save it

If we're to play our part in reducing the dangers of climate change, we need to make serious cuts in our energy use. But how can these cuts best be made? In Chapter 7 you'll find out what you can do as an individual, but here we look at the first big steps we can take as a society to save energy.

where the savings will come from

Imagine living in a home that was so well designed and insulated that it almost heated itself. Or living in a city where clean, fast public trains and buses turned up on time and took you where you needed to go. Where traffic jams, road rage and pollution were a thing of the past.

Making our energy work harder doesn't mean going back to the Dark Ages – or even to the winter of 1974 when a three-day working week, electricity rationing and power cuts got the country through an energy shortage. Although we use more energy now than we did in the 1970s, we can save more as well. In fact the UK could cut its energy use by nearly a third just by eliminating waste.

In 2007 the government outlined energy-saving measures that would reduce carbon emissions by 6 per cent by 2020, saying that two-thirds of those savings would be from households, a quarter from business and 10 per cent from the public sector. Critics said even more could be done.

reasons to use energy efficiently

Fewer carbon emissions = cleaner air
Fewer carbon emissions = better for the climate
Do more with less = save money, reduce waste

how can these cuts best be made

The main sources of carbon dioxide emissions in the UK are electricity generation, road transport, energy use in the home, and industry.

homes and buildings

There are 21 million homes in the UK, from modern apartments to 200-year-old houses. Most rely on electricity from the National Grid and use gas to provide hot water and heat, with some households using solid fuel or oil as heating fuel.

On average, each UK household uses energy that produces 6 tonnes of carbon dioxide a year. Typically, 30 per cent of our overall carbon footprint from homes comes from heating, a further 11 per cent from heating water, 13 per cent from electricity for lighting and appliances and 3 per cent comes from energy used for cooking.

Although we need to use energy to keep warm, cook and run the fridge, leaky buildings and inefficient appliances aren't making the best use of a lot of the energy we burn. In many UK houses almost half of all the energy used for heating simply escapes through the roof, windows and walls. Draughty buildings mean our central heating systems are working almost as hard to heat the outside as they are the inside of our homes.

But it doesn't have to be that way. Designers have come up with plans that could slash the amount of energy we use in our homes: common-sense measures such as insulating walls and loft, double-glazing and lagging hot water pipes can quickly lead to less wasted energy.

Some local authorities have been trail-blazing in the race to improve energy efficiency in the home. In 2003 when the London borough of Merton stated that new developments in the borough must get 10 per cent of their energy supply from renewable sources, builders responded by looking for ways to make houses more energy-efficient. They wanted to keep their costs low by shrinking the absolute size of that 10 per cent of renewables. More than 160 other local authorities have since followed Merton's lead.

The government has been criticised for failing to support the roll-out of the Merton Rule nationwide. To reach zero-carbon status, a building must have no overall carbon emissions as a result of energy use – achieved through high standards of insulation and heating design, combined with renewable sources of energy.

There are encouraging signs that the move to better building design and efficiency is taking hold – especially in continental Europe; but even around the UK there are a growing number of low-carbon buildings. Some individuals are achieving remarkable savings by adapting or building their own homes; and some commercial developers are beginning to see that energy-efficient buildings will make good business sense.

To encourage buyers to go for low-energy housing, zero-carbon homes are exempt from Stamp Duty (about £2,100 for the average house). An estimated 9 million new homes are needed in the UK by 2050: with that in mind, tougher building standards would do a lot to reduce the UK's carbon footprint.

fact

Building standards for new and refurbished properties are set out by the Building Research Establishment (BRE). A property built to its 'very good' Eco Homes standard will cut carbon emissions by 32 per cent, as well as reducing water use by 29 per cent

Where heat is lost from a typical house

(% of total heat loss)

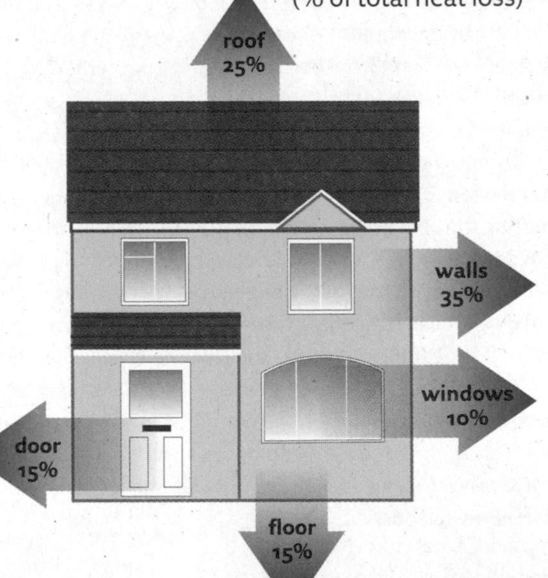

smartening up existing properties

Most of us don't live in new-build energy-efficient homes. Britain has streets of draughty Victorian and post-war properties – more than 4.4 million homes were built before 1920 and another 3.6 million date from pre-1945. Even by 2050 we're more than likely to be living in a building that's already standing today. So there's no sense pinning all our hopes on new buildings. We need to deal with the big carbon footprint of the buildings we already have. What can be done to make them more energy-efficient?

The government estimates that what it calls cost-effective energy-saving measures could reduce carbon emissions by 17 per cent by 2050. Adding loft insulation to a house can save almost 1.5 tonnes of carbon dioxide a year, while cavity wall insulation can reduce emissions by a further 1 tonne a year – saving up to £160 of household energy bills into the bargain. Millions of homes in the UK could benefit from cavity wall insulation.

Older properties can be insulated internally and externally. Internal insulation can save more than 2 tonnes of carbon dioxide a year. External insulation costs more and can be difficult to install if there is not enough space under the eaves of the roof. Savings of up to 2.5 tonnes of carbon dioxide a year are possible for a typical three-bedroomed semi-detached house.

clever ways to save energy

Energy-efficient appliances inside the home can help lighten the carbon footprint of daily life. European standards have encouraged manufacturers to produce smarter fridges and washing machines, but electricity use in the home is still increasing because of the burgeoning number – and in some cases their increasing sophistication. On the other hand, clever gadgets can reduce the amount of electricity appliances use. Smart plugs and remote controls can cut emissions, simply by eliminating the use of Standby and intelligent central controls can improve the efficiency of your heating system. So-called 'optimum start technology', another way of saving energy, works by reducing the need to put heating on in advance.

Freiburg – Germany's solar city

In 1944 British aircraft flattened more than 80 per cent of Freiburg in the south of Germany. Walk around the city today and you see an astonishing transformation. Not only have many historic buildings been restored but Freiburg has become a beacon of sustainability.

From its railway station to football stadium, this city of 200,000 people has almost as many solar panels as the whole of the UK. Thanks to a feed-in tariff the electricity companies buy solar power generated by households at up to three times the price that the householders buy electricity. This has made solar electricity cost-effective – and the impact is clear.

Solar panels producing both electricity and hot water are everywhere you look. Freiburg is also home to Europe's foremost solar power research institute and a host of solar-powered concept buildings. One of these is a house that rotates on a pillar, following the sun. Another is an office tower with the southern side completely covered in solar panels.

The Freiburg suburb of Vaubon gets two thirds of its electricity from the sun. Its 'passive houses' virtually heat and cool themselves. Built on a North-South orientation to make the most of sunlight, with all windows double-glazed and large south-facing ones to let in more sunlight, they are also highly insulated to keep in the warmth. Fresh air is brought in through underground ducts using the soil to regulate

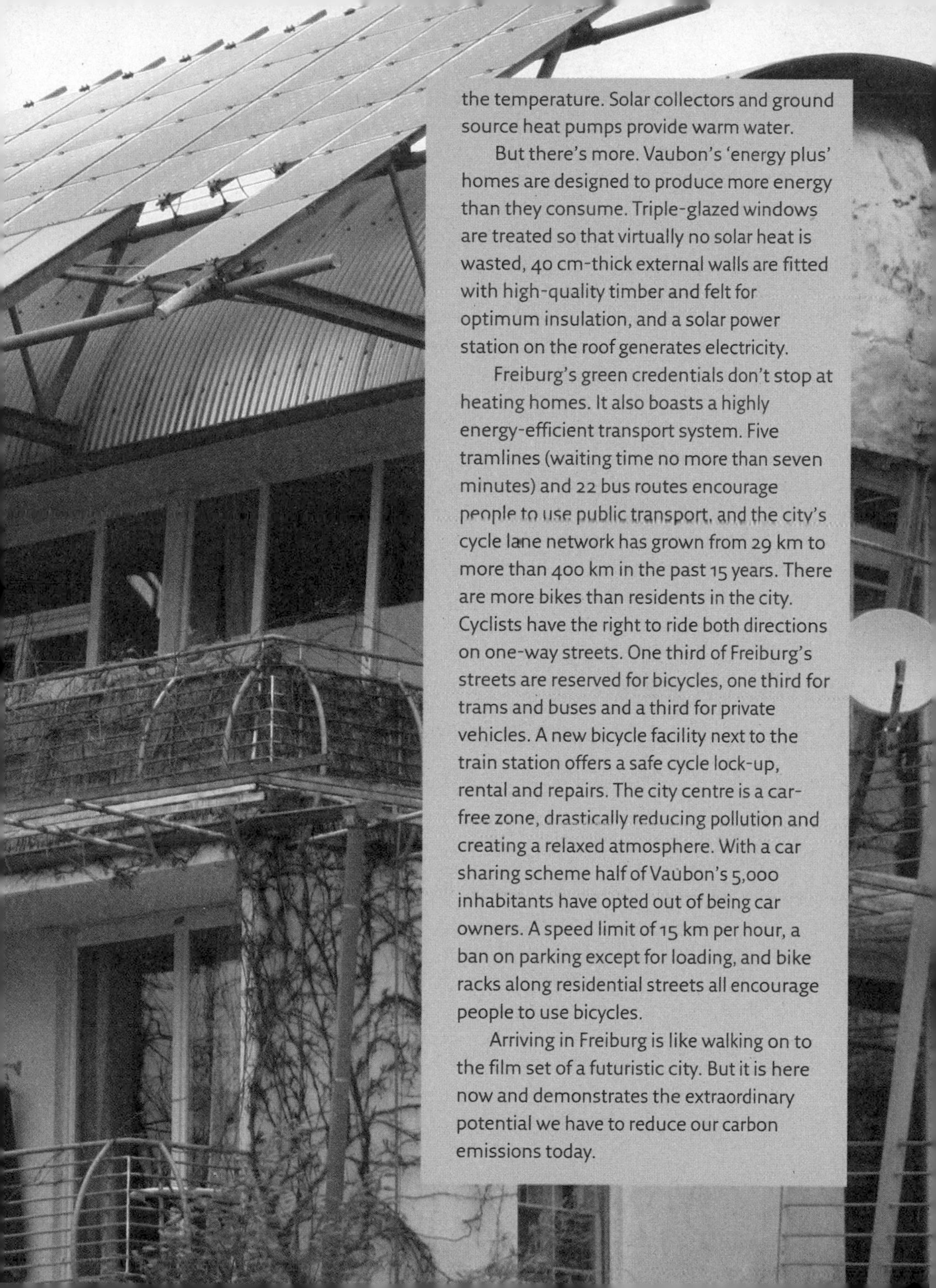

the temperature. Solar collectors and ground source heat pumps provide warm water.

But there's more. Vaubon's 'energy plus' homes are designed to produce more energy than they consume. Triple-glazed windows are treated so that virtually no solar heat is wasted, 40 cm-thick external walls are fitted with high-quality timber and felt for optimum insulation, and a solar power station on the roof generates electricity.

Freiburg's green credentials don't stop at heating homes. It also boasts a highly energy-efficient transport system. Five tramlines (waiting time no more than seven minutes) and 22 bus routes encourage people to use public transport, and the city's cycle lane network has grown from 29 km to more than 400 km in the past 15 years. There are more bikes than residents in the city. Cyclists have the right to ride both directions on one-way streets. One third of Freiburg's streets are reserved for bicycles, one third for trams and buses and a third for private vehicles. A new bicycle facility next to the train station offers a safe cycle lock-up, rental and repairs. The city centre is a car-free zone, drastically reducing pollution and creating a relaxed atmosphere. With a car sharing scheme half of Vaubon's 5,000 inhabitants have opted out of being car owners. A speed limit of 15 km per hour, a ban on parking except for loading, and bike racks along residential streets all encourage people to use bicycles.

Arriving in Freiburg is like walking on to the film set of a futuristic city. But it is here now and demonstrates the extraordinary potential we have to reduce our carbon emissions today.

Changing light bulbs can yield yet more savings – in fact the International Energy Agency estimates that we could save 10 per cent of our global electricity bill by switching to energy-saving light bulbs. Each 100 W bulb replaced by a compact fluorescent light (CFL) bulb will save up to 38 kg of carbon dioxide a year – and will last 12 times as long. Incandescent tungsten filament light bulbs are to be phased out by 2011 and LEDs (light-emitting diodes) could become 80 per cent efficient in the future. Simply switching off the lights can make a difference too. Research has shown that we waste £180 million worth of electricity by leaving lights switched on – and 770,000 tonnes of carbon dioxide a year.

If we were more aware of how much energy we were using, the chances are we would use far less. So, rather than hiding the electricity meter away under the stairs, putting it on display can help save energy, cutting electricity consumption by between 5 and 15 per cent. Households can install free meters as part of a government-backed trial and there's an increasing number of canny gadgets to keep you on top of your home's electricity use.

top tip

The Energy Performance Certificate (issued by an Energy Assessor) is a major part of the new Home Information Packs (HIPs) which are now required in England and Wales when you sell your home (right). The certificate shows your home's energy efficiency rating from A (most efficient) to G (least), and its environmental impact (CO_2) rating as well as advising how both of these can be improved.

save to spend

It's a curious thing: although there's huge potential for saving energy in our homes, offices and industry, evidence suggests that energy savings can, perversely, lead to more energy use.

A theory known as the Khazzoom-Brookes Postulate describes how, in a free market, increased energy efficiency results in cheaper prices – so consumption increases. Add to this the fact that companies take advantage of savings by spending the money elsewhere. Halve your gas bill and you have extra cash to spend on other things, which are likely to have their own carbon price. This is known as the rebound effect.

Energy efficiency is good for householders and for business. It means that more can be done with less. It also means that people who previously could not afford to have warm homes, may now be able to. But if the end result is an increasing demand for energy, this could mean more carbon dioxide emissions.

Not everyone agrees with the Khazzoom-Brookes Postulate, and there is evidence to suggest that the rebound effect is small compared to the total amount of energy and emissions saved. But the theory sounds a warning. Being more efficient is the first, necessary step. But we need to make sure the savings are invested in further efficiencies.

Energy Performance Certificate

17 Any Street,	Dwelling type:	Detached house
Any Town,	Date of assessment:	02 February 2007
County,	Date of certificate:	[dd mmmm yyyy]
YY3 5XX	Reference number:	0000-0000-0000-0000-0000
	Total floor area:	166 m²

This home's performance is rated in terms of the energy use per square metre of floor area, energy efficiency based on fuel costs and environmental impact based on carbon dioxide (CO_2) emissions.

Energy Efficiency Rating

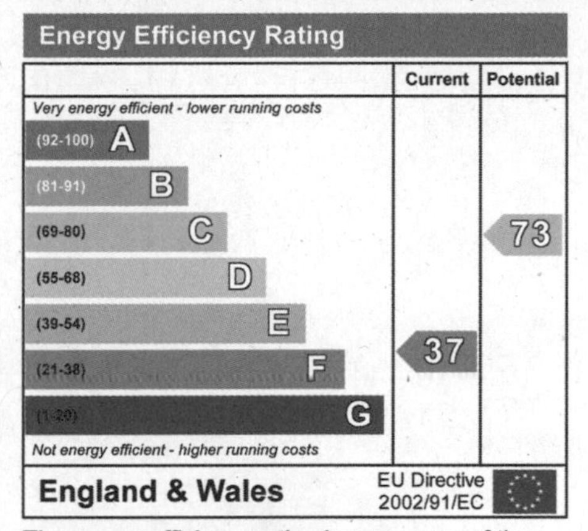

The energy efficiency rating is a measure of the overall efficiency of a home. The higher the rating the more energy efficient the home is and the lower the fuel bills will be.

Environmental Impact (CO_2) Rating

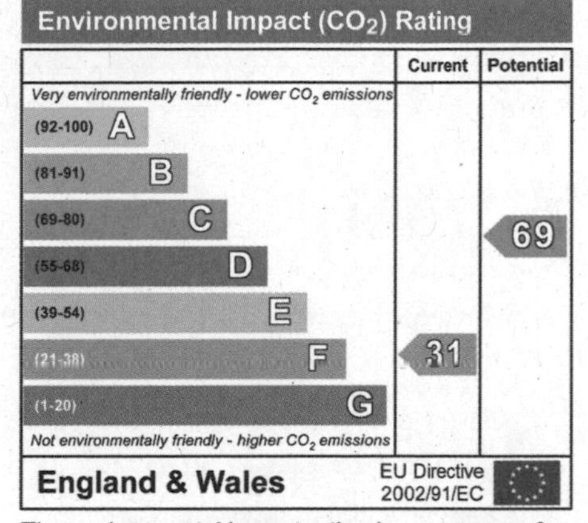

The environmental impact rating is a measure of a home's impact on the environment in terms of carbon dioxide (CO_2) emissions. The higher the rating the less impact it has on the environment.

Estimated energy use, carbon dioxide (CO_2) emissions and fuel costs of this home

	Current	Potential
Energy Use	453 kWh/m² per year	178 kWh/m² per year
Carbon dioxide emissions	13 tonnes per year	4.9 tonnes per year
Lighting	£81 per year	£65 per year
Heating	£1173 per year	£457 per year
Hot water	£219 per year	£104 per year

Based on standardised assumptions about occupancy, heating patterns and geographical location, the above table provides an indication of how much it will cost to provide lighting, heating and hot water to this home. The fuel costs only take into account the cost of fuel and not any associated service, maintenance or safety inspection. This certificate has been provided for comparative purposes only and enables one home to be compared with another. Always check the date the certificate was issued, because fuel prices can increase over time and energy saving recommendations will evolve.

To see how this home can achieve its potential rating please see the recommended measures.

Remember to look for the energy saving recommended logo when buying energy-efficient products. It's a quick and easy way to identify the most energy-efficient products on the market.

For advice on how to take action and to find out about offers available to help make your home more energy efficient, call **0800 512 012** or visit **www.energysavingtrust.org.uk/myhome**

the art of the possible – the energy-saving house

Six low-energy homes are being built at Nottingham University as part of a teaching project run by the university's School of the Built Environment.

Reducing energy demand was the starting point for the houses. But it was also crucial, for the purposes of the demonstration, that they did not cost more than conventional properties.

Five of the six houses were designed with different building methods and the latest energy-saving technologies – an upgrade of a 1930s house would complete the project. When the homes are finished, students or staff of the university will live in them, and their energy use will be monitored continually via a website.

Dr Mark Gillott, project leader, explains that making the homes desirable was crucial: 'We wanted to give the houses a wow factor when people walk in. We want people to think "I want to live in this house."'

Innovative building techniques are key to making the homes affordable. A four-bedroomed family house – the first property built – uses a lightweight recycled steel frame, rather than bricks and mortar, but with its pitched roof, looks like a conventional home. The frame, which took less than five days to put up, is insulated with plasterboard designed to absorb and give out heat. Savings in construction time and labour costs meant more money could be spent on the latest technologies to generate and conserve energy.

The foundations use an insulated concrete framework; but rather than conventional concrete this is made primarily from ground granulated blast furnace slag – a byproduct of the iron industry, which produces a tenth of the carbon dioxide emissions produced in the

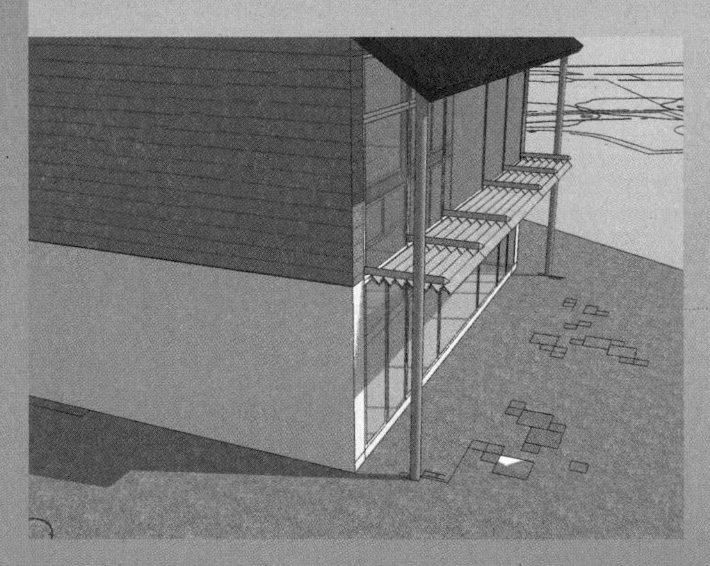

manufacture of Portland cement. Good design and renewable technologies combine to provide power and heat. An underground pump channels warm air (11–12°C) into the south-facing conservatory on the ground floor of the house, where it is warmed by the sun, and then pumped into the rest of the house.

In the loft a ventilation and heat-recovery unit converts solar energy from roof-mounted panels to produce hot water and warm air for under-floor heating. South-facing windows are triple-glazed and have blinds and external louvres to stop it getting too hot in summer. Balconies add to natural shade.

Smart energy controls allow the house to respond to changes in the weather – or to be controlled remotely via a website or a phone line. Appliances are all energy-saving models and lighting uses highly efficient light-emitting diodes (LEDs).

A water-saving system uses rainwater for the outside tap, washing machine and downstairs shower; and water from the bath and shower for flushing toilets. Aerators in the shower fittings increase the water pressure, and sensors on tap fittings ensure they are only used when needed.

Dr Gillott says, 'It's designed so that there's plenty of daylight, and it's a warm and comfortable place to be in winter and ventilated and airy in summer.' Good design, he emphasises, is crucial. 'It's what we say to the architecture students all the time. Don't leave it to the engineers to bolt on solutions afterwards.'

Two of the five new-build properties will cost under £60,000 and Dr Gillott hopes one of these will achieve zero-carbon status. 'There are no barriers from the technology to say that we cannot generate all the energy we need. There's no reason why we can't generate a zero-carbon home,' he says. 'The question is whether the industry will embrace it quickly enough to achieve it. Really, it is only legislation that can make the difference.'

the workplace

Just as big energy savings could be made in the housing sector, we can also smarten up the carbon impact of the workplace – whether that's a major corporation, government building or a home study.

Energy-efficient buildings play an important role. Good design can, for example, make the most of natural light and provide shade from too much sun. Here are three examples:

A recent Royal Navy building in Arbroath uses thermal heat stores and sunpipes to make the most of natural warmth and daylight in an energy-efficient office building for the Royal Marines.

The Swiss Re building, affectionately known as the gherkin, was designed as 'London's first environmentally progressive working environment', making use of natural light and ventilation, with an extracted air cooling system.

A Tesco store at Diss in Norfolk is designed to use 20 per cent less energy. Small steps such as adding doors to chill display units can make a huge difference to a supermarket's energy use.

For what you can do in the workplace, turn to Chapter 9.

PRETTY PICKLE:
London's iconic Gherkin reveals how architects can combine cutting-edge design with environmentally advanced features to reduce the carbon footprints of office buildings.

Making the scale of change needed to tackle climate change will take more than a handful of companies doing their bit. Good news, then, that one of the UK's biggest employers, the National Health Service (NHS), is trying to get to grips with its energy use. The NHS employs more than 1 million people and estimates suggest it emits 1 million tonnes of carbon a year. Hospitals have been set targets for reducing emissions and the Carbon Trust (a government-funded independent company that helps businesses and the public sector to do this) is working with hospitals and primary care trusts to find the most cost-effective ways of achieving their emission targets. Tom Cumberlege, Public Sector Manager at the Carbon Trust, says part of the challenge is to get all the staff involved, rather than leaving energy management to the estate manager. As with homes, schools and commercial buildings, the staff at hospitals who are there day in day out can make a big difference to their impact on the environment.

changing behaviour, cutting emissions

Changing the way people behave is one way of cutting emissions in the workplace, according to Chris Large. Chris manages Global Action Plan's Environment Champions programme, working with businesses to reduce their impact on the environment. He recruits volunteer Champions from across an organisation to help identify how changes can be made. The Champions come up with ways of persuading their colleagues to save energy, recycle more and use fewer resources. They audit environmental impacts across the business, looking at energy use, lights left on, waste and recycling rates. 'Champions think of things that people can do to make a difference,' Chris explains. 'It can be really simple guidance, such as how to print double-sided, or how to spell-check before your print.'

Some offices have introduced rewards for people who turn off their computers at the end of the day; others issue parking-ticket style warnings to those who forget. Champions also work on ways to get their message across. Knowing your office throws away a stack of paper higher than Big Ben can motivate people to think before they print, Chris says.

Three to four months into the project, the Champions carry out a follow-up audit to see what they have achieved.

And the results? Staff at the Britannia Building Society's London head office, to take one example, boosted paper recycling by 59 per cent; in the Leek office, staff saved £3,000 on fuel bills. A survey of employees' attitudes at Britannia's Leek office found that nearly three-quarters claimed to have changed their habits at work as a result of the scheme, and nearly a third said they had also changed what they do at home.

a degree cooler - London School of Economics

The London School of Economics and Political Science (LSE) is one of 48 universities so far taking part in a government-sponsored effort to curb the carbon impact of getting a degree in the UK.

With more than 170,000 square metres of floor space and 40 properties around the capital the LSE is energy-hungry: in 2006/07 its energy budget was some £2.5 million for gas, oil and electricity. The use of renewable electricity in many buildings has already reduced the greenhouse gas emissions significantly to 6,920 tonnes in 2007 and numerous energy efficiency projects underway which shave another 1,000 tonnes off that.

The Higher Education Carbon Management Scheme, supported by the Carbon Trust, offers institutions advice on cutting waste, energy, emissions - and bills. And it seems to be working: the universities taking part in 2006 generated a saving of 55,000 tonnes of carbon dioxide and £3 million.

LSE's Environmental Manager, Victoria Hands says: 'The School is looking at all areas of activity and involving a broad range of stakeholders to reduce carbon dioxide emissions. Its new Academic Building - designed to achieve an 'excellent' rating by the Building Research Establishment's environmental method (BREEAM) - incorporates a number of energy-saving features. For starters they didn't simply bulldoze the old building and start again, with all the new materials and waste that would imply: parts of the old building have been re-used with the foundation, the façade and the structure reintegrated into the design. A ground-source pump extracts cold water from an aquifer approximately 75 metres deep to provide comfort cooling to teaching rooms and lecture theatres. The basement houses recycling facilities and secure bicycle parking, showers and lockers. A highly efficient boiler has replaced the 1980s model, making a significant impact on energy consumption. Elongated windows on the lower floors and an atrium allow lots of daylight in, reducing the need for electric light, which is further controlled by movement detectors. Offices and seminar rooms have adjustable fan units which may be turned off if windows are open. Lecture theatres and classrooms have air quality and temperature sensors to regulate heating or cooling. Solar thermal collectors on the roof will supplement hot water requirements. The roof has also been strengthened to support two wind turbines that may be installed with the appropriate planning permission. A rooftop garden attracts birds and insects.

LSE has adopted sustainable and renewable energy measures throughout its buildings. New monitoring systems regulate indoor temperatures providing heating and cooling in the most efficient way. These systems are kept at a minimum during holidays. A software programme controls energy consumption in many buildings and halls of residence.

Director of planning and development, Julian Robinson, says he's proud to be working for an organisation that is serious about its wider environmental obligations. He sees the new academic building setting a benchmark 'which we will expect to exceed on our next major building project'.

a healthy approach to cutting emissions

Guy's and St Thomas' NHS Foundation Trust in London spends more than £10 million a year on energy – so finding ways to cut consumption made financial sense as well as helping to cut carbon emissions.

Working with the Carbon Trust, the two hospitals, which treat 750,000 patients every year, aim to knock a fifth off their carbon emissions, with savings of 14,000 tonnes of carbon dioxide already identified.

Good housekeeping and better energy awareness among the 9,500 staff will help reduce energy use, says David Porter, Head of Estates Management at Guy's and St Thomas'. But the Trust is also investing in improved technology to boost efficiency.

Staff are encouraged to switch off lights, recycle more and think about energy use as they go about their jobs, with posters around the sites and energy-saving messages on the Trust's intranet. Staff energy reps meet regularly to come up with new ideas.

A £2 million programme is upgrading lighting controls, improving insulation and fitting thermostatic valves on radiators in the hospitals, with estimated annual savings of £1 million. Maintenance staff are looking at how the hospital buildings can be run more efficiently, ensuring that ventilation systems are turned down when units are not in use. Plans to install combined heat and power on both sites will allow the hospitals to generate some of their own electricity, reducing energy bills and carbon emissions significantly.

David Porter says, 'Everyone can make a contribution, no matter how small each individual action appears. Small changes can add up to significant amounts, both in terms of financial savings and environmental benefits.'

industry

Industry accounts for a big chunk (18 per cent) of the UK's carbon dioxide emissions; and electricity-generating power stations are responsible for a further 27 per cent. Between these two sectors, nearly half the UK's emissions are outside the direct control of most of us.

For more on how we need to clean up the electricity sector, see Chapter 5.

Economic experts at the IPCC say that setting a high price for carbon would make it more attractive for industry to be more efficient. Energy-efficiency measures could create savings of 7-10 per cent without a carbon price, increasing to 23-46 per cent savings if the price of a tonne of carbon rose to US $100.

In the UK the government has so far used a number of financial regulations and incentives to get industry to change. It has sought to encourage action to cut emissions, providing advice and guidance through the Carbon Trust and Envirowise (see p.390). It has introduced a special tax – the Climate Change Levy – on the energy that business and public sector uses. It has also backed emissions trading as a means of encouraging energy efficiency, with the UK part of the Europe-wide Emissions Trading Scheme (see Chapter 6). Originally designed for large energy users, Emissions Trading has subsequently been extended – so that more companies will have to buy the right to pollute.

business savings

Chemicals manufacturer Holliday Pigments in Hull found it could save energy by using some of the heat generated by processes in its factory to generate steam, saving some £50,000 from energy bills and reducing its carbon footprint. Broadcaster Sky cut its footprint by 5 per cent in 2005-2006 through steps such as improving energy efficiency in its buildings and vehicle fleet. It also set up a carbon credit card for staff, giving points for carbon-friendly behaviour such as cycling to work. The company switched to renewable electricity supplies and offset its remaining emissions by investing in renewable energy projects in New Zealand and Bulgaria.

offsetting – does it count?

Carbon offsetting has become big business in recent years with companies offering to offset emissions by investing in carbon-saving projects. The idea is that for every tonne of carbon dioxide produced, a tonne is either removed from the atmosphere – by planting trees intended to act as a carbon sink – or avoided, by replacing a polluting activity with a clean alternative, such as renewable energy.

Some offsetting schemes have been criticised on a number of counts. Although trees remove carbon dioxide from the atmosphere, they must remain *in situ* to prevent the carbon being released again, so there would have to be strong guarantees that forests planted this year to offset emissions were not cut down a couple of years later. There are other problems with tree-planting offset schemes, too: large plantations where just one type of tree is grown are bad news for wildlife and can cause problems for people locally.

Offsetting schemes that invest in renewable energy projects can benefit communities in the developing world, and a gold standard has been established for offsetting schemes that deliver genuine benefits. This guarantees that the investment is made in a scheme that would have struggled to find funding in any other way.

However robust an individual offset scheme, Friends of the Earth points out that it is simply a way for the industrialised world to buy the right to continue polluting rather than cutting emissions at source. Offsetting allows companies and individuals to claim that they are tackling carbon emissions when they may not be doing anything to cut emissions at source. Such an approach will do little to reduce the UK's emissions of carbon dioxide. The spread of carbon offsetting may have helped stimulate awareness of climate change, but many people now argue that it should be a last resort – and that genuine cuts to carbon emissions must come first.

local authorities

The government estimates local authorities can cut their own energy use by 11 per cent. Although local authorities themselves are responsible for a small proportion of UK emissions they have powers beyond their own buildings, land and work force: they also influence suppliers and services through the goods and services they buy; they control local planning and development, and are responsible for green spaces and promoting public transport.

A scattering of local authorities have embraced the need to take action on climate change. More than 220 have signed a commitment to make this a cornerstone of their policies. The year 2000 saw the launch of The 'Nottingham Declaration on Climate Change' (updated in 2005), in which all signed up councils were encouraged to develop a climate change action plan and set targets for reducing emissions.

Nottingham City Council, the first authority to sign, has reduced its carbon emissions by 30,000 tonnes by using renewable energy supplies. Its policies have helped stabilise emissions from transport in the city and boosted recycling rates. The council is now looking to become carbon neutral.

schools

Some 6,000 primary and secondary schools across the UK are involved in the Eco-School scheme, which encourages teachers and pupils to work together to come up with ways to save energy, improve recycling and save water at school.

As well as cutting emissions, schools taking part find the scheme has other benefits, including better links to the wider community, improvements in the school ethos and relations between staff and pupils, and reduced bills. Some funding is available from the government to support low-carbon and energy-saving measures, but critics say there is not nearly enough to go round.

transport

Road transport accounts for between a fifth and a quarter of UK emissions. Emissions from transport are on the up: more and more vehicles on the road mean that carbon dioxide emissions have risen by 9 per cent since 1990. With government figures showing carbon dioxide emissions from transport set to soar, can anything be done to save energy and emissions in the transport sector?

According to the experts, savings are possible but require a big shift in behaviour, encouraged by changes to transport policies – locally and at national level.

on the road

When it comes to travelling by car, there are two basic ways to reduce the level of emissions: driving more efficiently and driving less.

driving more efficiently

More efficient driving means getting more miles out of every gallon of fuel or using different fuels. So we need to look at the way people drive and the types of car they use. The performance of modern vehicles varies enormously, from small vehicles that can do 65-70 mpg, to big gas-guzzlers that manage less than 15 mpg. Carbon dioxide emissions are directly related to miles per gallon and so can vary from 104 g/km for a Toyota Prius hybrid to 520 g/km for a sports car (Lamborghini Diablo 132) and 387 g/km for a 4X4 (Toyota Land Cruiser). Within each class of car performance varies too: at 186-219 g/km, for example, a Lexus hybrid is much more polluting than the Prius. An average motor emits 90 kg of carbon dioxide for every full tank of petrol used.

A UK government-commissioned study found that, assuming the car will continue to be the way most people choose to get around, the best thing to do will be to make cars as low-carbon as possible. The European Union has set a target for car manufacturers to reduce average emissions from new cars to 140 g/km carbon dioxide by 2010; new cars sold in the UK in 2006 were above this average, at 167.2 g/km. Tougher emissions standards for vehicles would make one of the biggest contributions to cutting transport's carbon fuel bill.

driving less

Another way of cutting emissions is to reduce the numbers of car journeys. In the UK 17 million people go to work by car. Workplace travel schemes can make it easier to leave the car at home – with employers providing cycle facilities, showers and loans to buy a bike, or setting up lift-share schemes or loans for season tickets on public transport.

SUPER CYCLEWAYS:

By 2025 London could be the cycling capital of the world. Around £4 million will see cycling become a fully-funded part of the public transport network. Plans include a new free bike lending scheme (offering 6,000 bikes every 300 m) and 12 cycle highways to make pedal power a real travel choice for everyone.

Flexible working can reduce the need to commute – working from home one day a week can cut transport emissions by 20 per cent. A study by BT into the impact of home working found that, on average, employees working from home reduced their carbon dioxide emissions by 15.2 kg per week. Even assuming that some savings would be lost by increases in home energy use, the study found total emissions for 5,000 employees were reduced by some 3,663 tonnes per year.

Cycling and walking can be easy alternatives for short car trips, but longer journeys and city-wide travel need better public transport. Good bus services, light rail and trams can provide alternatives to driving, substantially reducing carbon dioxide emissions.

The city of Curitiba in Brazil has been held up as a model of what public transport can achieve if investment and urban planning keep the needs of people in mind. The city, which has a growing population of more than 1.6 million people, invested in buses as its main public transport, creating special bus-only avenues. The system, which is cheap to use, carries some 2.14 million passengers a day despite high levels of car ownership. The result has been cleaner city air and some of lowest rates of fuel consumption per person in Brazil.

Freiburg in Germany (see also pp.94–95) is one of many European cities to have developed a low-carbon approach to getting about. The city has 160 km of sign-posted cycle paths, cycle-only streets and special access on one-way routes. Buses and trams carry 67 million people a year, with cheap tickets for families.

In France the Mayor of Paris has introduced a low-cost bike rental scheme in the capital, providing bicycles at 750 stations around the city. Users of the Velib bike scheme can buy either an annual pass (priced at £20) or a one-day pass (less than £1), with short journeys completely free. Paris has some 230 miles of cycle paths and the number of cyclists in the city has increased by 50 per cent in the past ten years.

A charge on drivers entering the city centre, combined with investment in public transport, has cut the level of traffic in inner London, with increases in the numbers of people travelling by tube and bus. One opinion survey found that 1 in 5 drivers would leave their car at home and use public transport if road charging schemes were introduced across the UK.

Improvements to public transport can tempt drivers out of their cars for longer distance trips, with both coach and rail travel offering lower carbon dioxide emissions per passenger mile. Planning policies can also have an impact – for instance local development plans can ensure new shops are accessible by public transport or by foot.

A UK government study found that with the right policies, emissions from road transport could be reduced by 60 per cent from 1990 levels by 2030. Some of these savings depend on using hybrid vehicles and alternative fuels, but measures such as road pricing, energy-efficient driving and changes to freight distribution systems also had a role to play.

VIVE LA REVOLUTION:

'This is about revolutionising urban culture,' says Pierre Aidenbaum, mayor of Paris' third arrondissement, as the capital provides more than 10,000 bicycles for Parisians and visitors alike.

potential emissions savings from different transport measures

measure	saving*
Low emissions vehicles	18.3 - 9.1
Alternative fuels	9.1 - 1.8
Greener driving	4.6 - 2.5
Improved freight/ local production	2.5 - 0.7
Travel plans, car clubs, car sharing	2.4 - 0.9
Better planning	2.4 - 0.5
Road pricing	2.3 - 1.1
Use of IT	1.2 - 0.3
Rail for air	0.7 - 0.5

*million tonnes carbon per year

Further savings would be possible if we changed the way we live in our cities and towns: better inner city housing and pedestrian and cycle access to shops and services encourage people to use their cars less. When planners redeveloped the German capital of Berlin, they included a car-free zone in the centre as part of the city's commitment to cut emissions by 40 per cent by 2020. Homes in the car-free area are surrounded by pedestrian and cycle paths, but no parking spaces. Local shops and services mean residents do not need to use a car. A new building code now restricts parking for new buildings in the city.

Better public transport, towns designed for pedestrians, less congestion, less pollution, and fewer journeys by car could all improve our quality of life.

air travel

Air travel accounts for a growing percentage of the UK's carbon dioxide emissions. If current levels of growth continue, emissions from aviation could account for 80 per cent of the UK's entire carbon budget by 2050.

There is certainly scope to make aeroplanes more efficient. Improvements have been made since the first jets came into service in the 1960s – although early jets were much more polluting than their propellor-driven predecessors and jets have only recently reached the same level of efficiency as the engines of the 1950. Aircraft manufacturer Boeing advertises that its recent 747-8 Intercontinental aircraft has improved fuel efficiency by 15 per cent, resulting in 75 g per passenger km (that's equivalent to 202 g/km if we count the effect of emissions in the upper atmosphere). New designs could help improve aircraft efficiency. The flying wing – or blended wing-body design, which extends the passenger cabin into the wings of the aircraft, could make planes lighter and more aerodynamic, so saving fuel. Its champions claim it could reduce emissions by 90 per cent within 50 years. Plans are still on the drawing board, however, and while engineers are confident that the planes could be built, they are not yet considered commercially viable.

Research by the Tyndall Centre for Climate Change Research suggests that for aviation to play its part in keeping carbon emissions within safe limits, it would need to make efficiency gains much faster than it has ever done in the past. In fact the Tyndall Centre's research suggests aircraft engine efficiency will improve by such a small amount each year that any gains would be more than cancelled out by the rapid growth in the number of flights we take. One problem with any efficiency improvements in aviation is that they can take such a long time to come into effect. The life span of the average passenger aircraft is some 30 to 50 years and two-thirds of all the aircraft likely to be flying in 2030 are already in use. Carbon cuts are needed long before that.

Aviation's growing share of the UK's projected carbon emissions

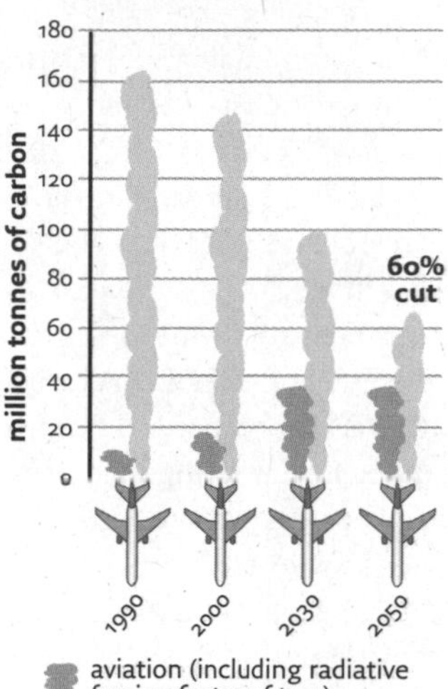

Future growth of air travel means its share of carbon emissions will become more and more damaging. Government projections for aviation's emissions may be too conservative – others suggest aviation's emissions in 2050 will make up far more of the UK's total.
Source: Draft Climate Change Bill (2007)

More immediate savings can be made by changing the way air traffic is managed – effectively reducing the time that an aircraft spends in the air. One assessment suggests this could reduce emissions by between 6 and 12 per cent.

The European Union intends to bring aviation within the Emissions Trading Scheme to provide an incentive for the aviation industry to reduce emissions. While some airlines have welcomed the proposal, modelling suggests it is unlikely to result in lower emissions as it is unlikely to have a major impact on ticket prices and passenger numbers.

Indeed, it is hard to see how savings from efficiency can compensate for the rapid growth in emissions from flights. Experts at the IPCC concluded that while savings are possible, these will only partly offset the growth in aviation emissions globally. Offsetting schemes may make us feel better as passengers, but the potential to offset all our emissions simply does not exist.

One way forward would be for airlines to pay for the environmental damage they cause. Air passenger tax could be raised to cover the environmental costs and curb growth in demand. The need to restrain growth also places a huge questionmark over airport expansion plans in the UK. The only really effective way to cut emissions from the aviation sector is to manage the amount of flying that goes on – that doesn't mean we have to fly less than we do now, but it does imply not flying much more. It also means doing something about air-freight, for example through lowering consumption of imported goods.

way to go: Friends of the Earth's travel plan:

1. Better public transport services: more reliable, more efficient, more affordable
2. Safer streets for cycling and walking
3. Safe routes to school for all children
4. Services and facilities close to where people live and work so they don't need to drive
5. Lower speed limits: 20 mph in residential areas
6. More use of technology to cut the need to travel: working from home, video-conferencing
7. Pay-as-you-go road user charging
8. More funding for rail freight projects

Friends of the Earth wants to see money earmarked for new roads spent on public transport schemes that will reduce rising transport emissions.

farming

Globally, emissions from agriculture are estimated to account for some 30 per cent of all greenhouse gas emissions, including carbon dioxide emissions from farmland, fertiliser use and methane from animals. In Europe emissions from farming are relatively lower – accounting for around 11 per cent of total greenhouse gas emissions – and these levels are falling because of the use of less inorganic fertilisers, less cattle farming and better land-management. UK farming's main contribution to global warming is through the production of methane and nitrous oxide. Our farms are responsible for almost half of our methane emissions and two-thirds of emissions of nitrous oxide. Methane emissions from dairy cattle have fallen over the past 40 years, partly as a result of changes to diet. More intensive production methods (which have other environmental costs) have meant fewer cows are needed to produce the same amount of milk.

BODY BLOW:

Livestock generate more greenhouse gas emissions than transport – so eating meat less often could be as good for the planet as for our health.

Emissions from animal manure can be reduced by cutting livestock numbers and by converting the gas produced to biogas; this can then be used as a fuel. Because the system relies on animals being housed rather than allowed to graze, there are concerns over animal welfare. The plants are expensive but in the United States subsidies have encouraged farmers to invest.

There is huge potential for reducing emissions from farming globally, according to experts at the IPCC. As well as cutting emissions of methane and nitrous oxide, we could make better use of soils to absorb carbon dioxide. It is estimated that 13 million tonnes of carbon are lost from all UK soils each year. Changes to farming methods, such as using organic fertilisers that produce less nitrous oxide, or reducing the amount of soil tillage, can reduce this amount.

Farmers in parts of Africa use natural mulches that help to moderate soil temperature, suppress disease and pests, as well as keeping the soil moist and retaining carbon. Changes in diet, particularly eating less meat and dairy produce, could potentially lower emissions from farming by reducing the quantity of livestock farmed.

watching the waste line

Dealing with our rubbish contributes around 2 per cent to the UK's greenhouse gas total. Most of this comes from methane emissions from landfill. But our growing rubbish mountain also reflects rising consumption – and that means more energy being used to manufacture stuff we end up throwing away. Every year, the average dustbin contains enough discarded energy for 500 baths, 3,500 showers or 5,000 hours of television.

So curbing waste can reduce emissions, as can re-use and recycling. Recycling aluminium, for example, uses just 5 per cent of the energy it takes to make new aluminium – and produces only 5 per cent of the carbon dioxide emissions. The energy saved by recycling one glass bottle can power a 100-watt light bulb for almost an hour. According to the Waste and Resources Action Programme (WRAP), current levels of recycling save some 10–15 million tonnes of carbon dioxide equivalent a year – about the same as taking 3.5 million cars off the road. The government has set a target for recycling half of municipal waste by 2020 (from our current 28 per cent). Austria, by comparison, has already achieved a 64 per cent recycling rate.

Composting food and garden waste diverts these materials from landfill, and can also produce good-quality compost that can replace peat, reduce fertiliser use, regenerate soils and hold some carbon in the soils.

CAN-DO APPROACH:

In 1989 we recycled just two in every 100 aluminium cans. In 2007 it was nearer 50 in every hundred – with huge energy savings.

It is also possible to create 100 per cent renewable energy from waste. Research for Friends of the Earth has found that if household food waste was collected and processed in anaerobic digestion plants, it could be used to generate electricity, potentially contributing at least 0.4 per cent of electricity demand. More power could be generated if food waste from cafes, supermarkets and canteens were added.

This is not the same as burning waste in incinerators to produce energy. Incinerators are very inefficient energy generators and can actually emit more carbon dioxide than a coal-fired power station. Recycling as much as possible, degrading biodegradable materials, then putting them in landfill, is better for the climate than incineration, with or without recovering heat. What's more, once built, an incinerator needs a steady supply of rubbish, discouraging recycling and waste reduction.

Our consumption of products – wherever they are made – adds to our impact on the climate. This means that when we recycle in the UK we may be saving energy and emissions elsewhere in the world. For example, recycling aluminium in the UK will avoid emissions in the country where the aluminium would otherwise have been mined.

top ten tips for saving energy in the UK

1. A nationwide programme to insulate our homes
2. Tougher environmental standards for new homes
3. Tougher efficiency requirements on all new electrical goods
4. Tougher emissions standards for all new cars
5. Increased investment in public transport at local authority level
6. No new airport growth
7. Higher Air Passenger Departure tax
8. Increased recycling in the workplace
9. Shifting freight off the roads and onto rail
10. Eating less meat and dairy produce

Household and other municipal waste is only around 10 per cent of all waste. The rest is made up of big stuff like commercial, industrial construction and demolition waste. There's huge potential for preventing a lot of this, and for making sure that where waste is created it is recycled or composted. Friends of the Earth believes one effective way of ensuring that this happens would be a phased ban on landfill or incineration of any material that is reusable, recyclable or compostable.

Preventing waste in the first place is the best approach, followed by maximising re-use, recycling and composting. In the long term we should be getting rid of residual waste through better collection schemes and re-designing products to make them more recyclable or compostable. This approach will help minimise the climate impacts of the materials that we use and throw away.

low-carbon diets

When Dr Andrea Collins at the BRASS Research Centre in Cardiff worked out the impact of changes in diet on the average footprint, the findings were surprising. A typical vegetarian diet, replacing meat with cheese and dairy products, led to a 5.9 per cent reduction in the footprint. But, perhaps surprisingly, a diet that replaced high-carbon items like beef with lower-carbon products like pork reduced the overall footprint by much more – some 26 per cent. Dr Collins explains that because dairy farming is more carbon-intensive than raising pigs, the footprint of cheese and other dairy products is larger than for pork and bacon.

A breakdown of Cardiff's footprint found that nearly a quarter (23 per cent) of emissions came from food and drink, including consumption at home and in cafes and restaurants. An average resident consumed almost 143 kg of milk and dairy products every year, 196 kg of fruit and veg and 72 kg of meat.

The study found the most carbon-intensive areas of the typical diet and looked at how changing what people eat might reduce the city's footprint.

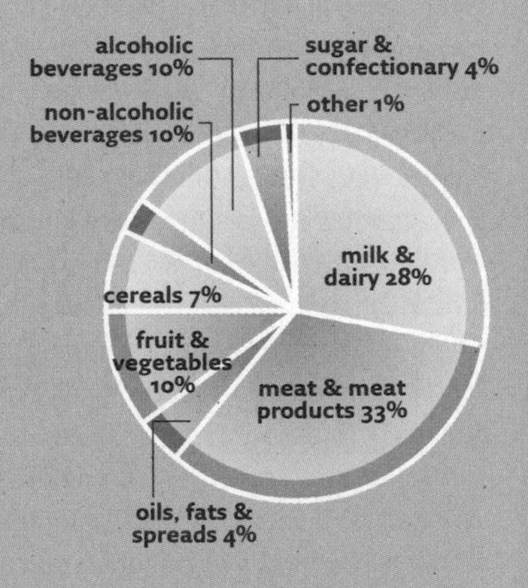

This breakdown of Cardiff's food footprint in global hectares per person (see p.75) shows the biggest impact is from meat and dairy products - foods we can cut down on with wins for health and the environment.

trees, soil and carbon stores

The way we use land has a big effect on the amount of carbon in the atmosphere. For example, forests, soils and peat all soak up carbon dioxide from the atmosphere and if they are managed well they can help us deal with carbon emissions. In the past few decades humankind has changed the land more quickly than ever before: more land has been converted to growing crops since 1945 than during the whole of the 18th and 19th centuries combined.

Most emissions from deforestation come from land converted to agricultural production: this process releases the carbon bound up in the trees and the soil. Dense tropical forests are rich stores of carbon dioxide, with 500-600 tonnes of carbon dioxide per hectare. In many tropical forests the land is being cleared for farming through a technique known as slash and burn. Burning is a cheap way of clearing land, leaving behind a layer of ash that provides fertiliser but it releases a lot of carbon dioxide. Western demand for agricultural produce – including beef and soy from the Amazon, and palm oil, coffee and timber from South East Asia – is driving the destruction of the forest. It's also a problem in parts of Africa where small-scale farmers clear trees to grow food to survive. Even in areas where the forests are legally protected, illegal logging can be a major threat, with consumer demand for wood and paper high.

Certification schemes such as FSC (run by the Forest Stewardship Council), which guarantees that wood and paper products have been sustainably sourced, can help support a more environment-friendly approach to the trade. This requires good forest management, with only a small number of mature trees felled, providing space for re-growth.

Small initiatives to protect the forest can make a difference, with local people often playing a crucial role. Recognising the land rights of people living in the forest is important: and the involvement of local people tends to encourage better care of the forests. Schemes that improve the potential for local employment can make it easier for people to survive without resorting to forest clearance. In the Philippines a switch to more

TIMBER:

Indonesia loses on average 2 million hectares of deforestation a year – that's equivalent to a tennis court for each person in the UK. Deforestation is largely the reason Indonesia is now the third largest emitter of greenhouse gases.

" The world's forests need to be seen for what they are, giant global utilities providing essential services to humanity on a vast scale.

HRH Prince Charles

labour-intensive rice-farming created jobs and halved forest clearance rates.

Some developing countries have suggested schemes that would allow them to be paid to avoid deforestation. In Costa Rica the government scheme pays landowners a fee per hectare to maintain their forests, storing the carbon and preserving wildlife habitat and water resources. Forest cover in Costa Rica has more than doubled since 1977.

Planting trees can help with the take-up of carbon dioxide. It can also help prevent soil erosion and improve drainage. But in areas where water resources are already stretched, the trees are unlikely to survive. New forests create pressures on land use, potentially competing with agriculture for fertile areas.

Reforestation projects are encouraged under the Kyoto Protocol – the international agreement to tackle climate change – whereby rich countries can claim carbon reductions by investing in planting schemes. But critics point out that the scheme actually encourages deforestation as the old is cut down to make way for the new. What's more, farming young trees provides only a short-term carbon store.

Changes in agricultural practices can reduce the level of emissions from farmland. Conservation tillage, which avoids breaking up the soil, improves fertility, which in turn helps retain carbon, as does setting land aside for grassland.

There are, of course, other pressures on land. In the UK, for example, there is an increasing move to put housing and other development on greenfield sites. Although land use in the UK does not have a major impact on global emissions of greenhouse gases, it can affect the way we deal with the effects of climate change. For example concreting over the land reduces natural drainage patterns, requiring more investment in drains and flood-management systems. Some 3.5 million hectares of land in the UK are protected from development, but just 10 per cent is now covered by woodland. Some of the UK's protected sites have suffered over the past century – bad news for the plants and wildlife that live there and for the climate.

Using natural carbon sinks is no substitute for cutting carbon emissions, but looking after the natural environment, from forests to peat bogs, can help keep the atmosphere clean.

firebreaking in the amazon

José Dolce is a farmer in Guarantã do Norte, in the southern part of the Brazilian Amazon. He has been farming in the area since the 1970s, but in 2000 was persuaded to change the way he farmed, thanks to a Friends of the Earth programme. 'Fire here used to be a kind of vicious circle,' he says. 'We have always been so used to fire, as if it were a part of our lives, that we did not really consider the alternatives.'

Friends of the Earth Amazonia has been encouraging farmers to stop using fire to clear their land, but changing years of farming practice is a challenge. 'Once we decided to try alternative forms of production and cattle raising, it was easier,' José concedes. 'We just used a system of rotation of pasture, by moving the cattle every few days and having natural regeneration in those areas where cows are not allowed.'

The results have been impressive, with improvements in the quality of the soil and the grazing pasture. Less burning also means a reduced risk of fire in the nearby forest reserve. 'It was like being born again,' José adds. 'The fresh humidity of the early morning is back, the respiratory illnesses have decreased, especially among kids and elder people.

'Now I am trying to make new experiments along ecological principles, and I started to produce organic watermelon. This is unusual in this region, but people seem to like it, they pay a good price and ask me to grow more.'

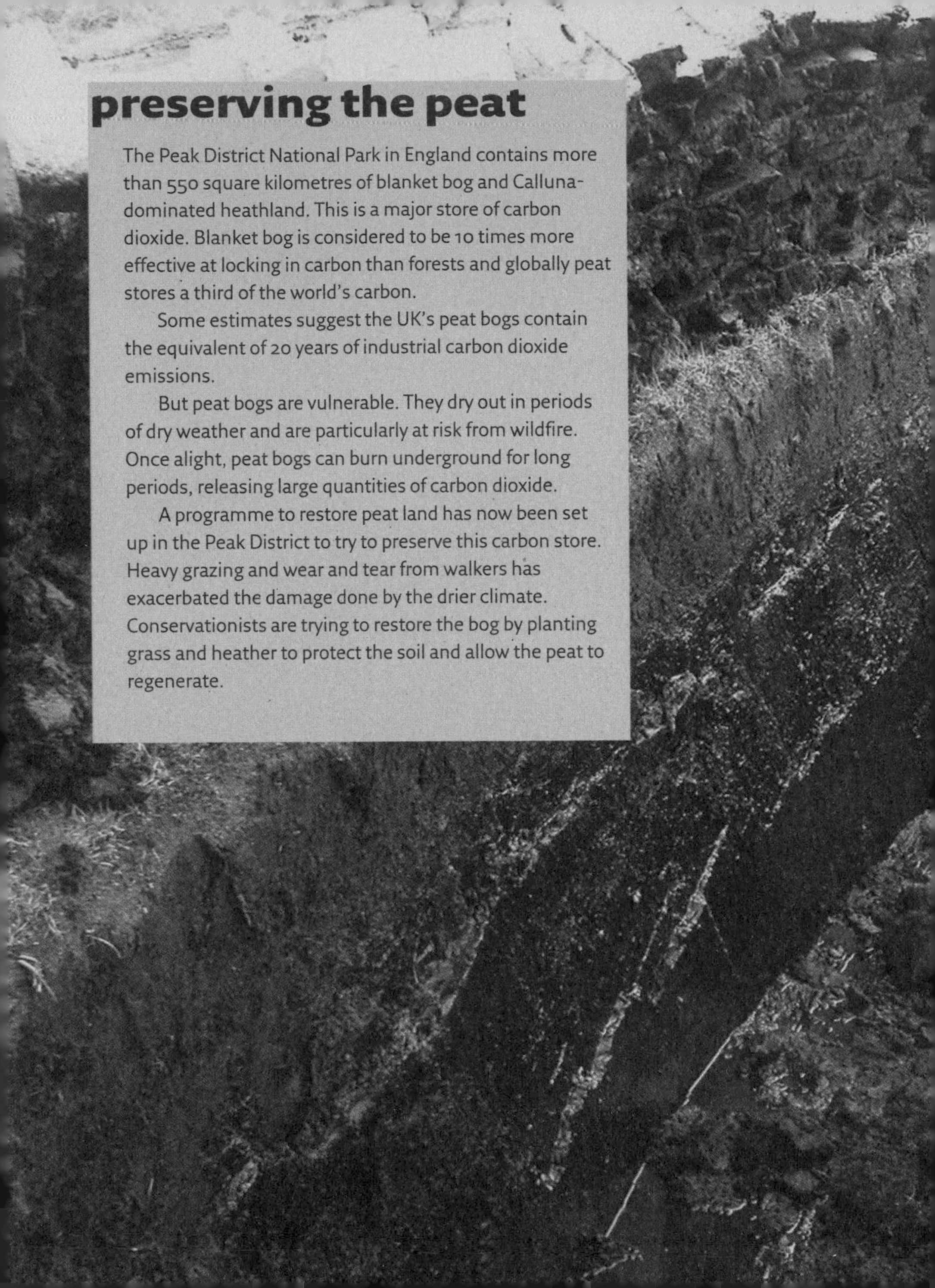

preserving the peat

The Peak District National Park in England contains more than 550 square kilometres of blanket bog and Calluna-dominated heathland. This is a major store of carbon dioxide. Blanket bog is considered to be 10 times more effective at locking in carbon than forests and globally peat stores a third of the world's carbon.

Some estimates suggest the UK's peat bogs contain the equivalent of 20 years of industrial carbon dioxide emissions.

But peat bogs are vulnerable. They dry out in periods of dry weather and are particularly at risk from wildfire. Once alight, peat bogs can burn underground for long periods, releasing large quantities of carbon dioxide.

A programme to restore peat land has now been set up in the Peak District to try to preserve this carbon store. Heavy grazing and wear and tear from walkers has exacerbated the damage done by the drier climate. Conservationists are trying to restore the bog by planting grass and heather to protect the soil and allow the peat to regenerate.

“ The way we manage our peat moorlands has a massive bearing on our ability to tackle climate change... It is the forgotten climate change timebomb.

Fiona Reynolds, Director-General, National Trust

cleaner
energy

chapter 5

Using cleaner energy is vital if we want to clean up our act. But what are the cleaner sources and how much energy can they supply?

looking at alternatives

Saving energy will make a big difference to the amount of carbon we pump into the air. But it's not enough on its own. We also need to re-think where we get our energy from. If we stick mainly to oil, coal and gas – which today supply most of our electricity, keep our transport running, and heat most of our homes and buildings – things will get worse. There are plenty of alternatives we could use that will have a much lower impact on the planet – and they are increasingly available. But how much energy can they supply? And what are their pros and cons? This chapter looks at the best ways to produce electricity and heat, and fuel our transport.

Over 90 per cent of the energy we use comes from gas, oil and coal. All these fossil fuels emit large amounts of carbon dioxide when burned, although there are differences in their efficiency: burning coal in a power station, for example, produces more carbon dioxide per unit of useable energy than burning petrol in a car.

The key thing to remember is that although electricity represents less than a fifth of all the energy we use, making it produces more than a third of UK's total carbon dioxide emissions. That's why we need to start by looking at alternative ways of generating electricity.

primary fuel source breakdown (2006)

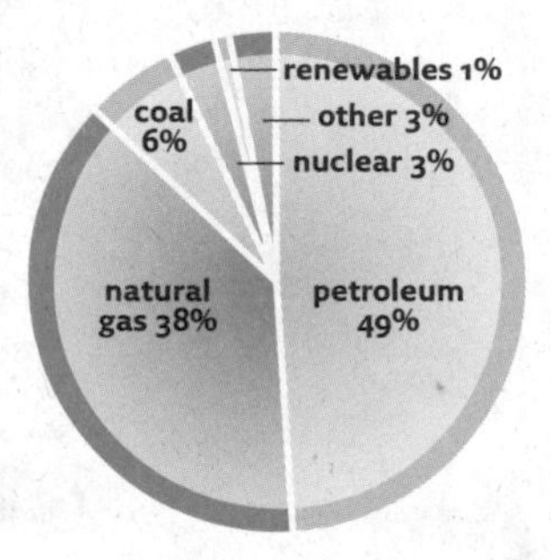

Final use for fuels in the UK (2006)

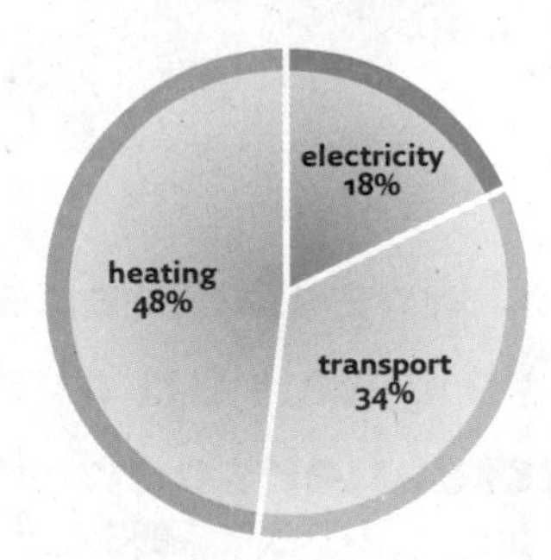

Source: BERR

measuring energy

- Primary fuels (coal, gas and oil) are often measured in tonnes of oil equivalent – i.e., the amount of oil that would yield the same amount of energy.
- Electricity is measured in watts, kilowatts (kW), megawatts (MW), gigawatts (GW) and terawatts (TW). A watt is a measurement of the current x voltage, indicating the amount of power generated.
- The electricity a power station produces is usually measured in megawatt hours (MWh); that is, the amount of energy generated over a period of time. Domestic electricity bills are generally measured in kilowatt hours (kWh).
- A 100 watt light bulb will use 100 watt hours, or 0.1 kWh, of electricity if left on for one hour.

generating electricity

Since producing electricity can be seen as the single largest source of climate-changing carbon dioxide emissions, let's start with how we keep the kettle boiling.

Most electricity used in the UK comes from big power stations burning coal and gas (and a small amount of oil). Coal tends to be the choice when gas prices are high – oil and gas usually get more expensive if there is a perceived threat to supply. Two things give the electricity sector a huge carbon footprint. First, our heavy reliance on fossil fuels; second, the fact that existing power stations waste about half the energy they produce – through heat that is simply released in cooling towers, and through losses during transmission down the cables. In effect we spend a lot of time and money heating up rivers and the sea. This is in part because our system for producing energy is highly centralised – in big power stations a long way from the consumer. Renewable low-carbon sources of energy such as wind power, hydroelectricity and biofuels as yet provide only a tiny proportion of our electricity. And we're not yet making nearly enough of the potential of decentralised energy systems – ie going off grid.

So how much electricity do we need? Consumption in the UK amounts to some 350 TWh a year. How much generating capacity will we need in future? This depends on how determined we are to manage demand. Government estimates suggest our demand for electricity will grow by 26 per cent by 2020. Modelling for Friends of the Earth suggests we could actually reduce consumption by a tenth by that date; other respected sources predict growth in demand of up to 1 per cent a year to 2020.

Some power plants will close by 2015. Almost all UK nuclear power stations are approaching the end of their life, while some coal-fired power stations are being phased out. At the same time North Sea gas is running low and the UK is increasingly dependent on gas imports. The government says the UK will need 20-25 GW of new generating capacity by 2020. So we have little option but to invest in new sources of energy. Even if we can curb energy use it's clear we'll need new capacity – and it will be much better for the planet if these are renewables.

CURRENT AFFAIRS:

In 2006 more than a third of the UK's electricity supply came from burning coal, slightly less from gas, less than a fifth from nuclear power and around 4 per cent from renewables.

how much renewable energy could Britain produce?

With some of the windiest weather in Europe and almost 8,000 miles of coastline, the UK is a power house waiting to be switched on. To date the main focus has been on wind power but wave and tidal energy are also coming through.

No single type of renewable energy is expected to power Britain. Visions of the landscape covered in wind turbines, or the coast entirely surrounded by wave machines are fantasy – or scaremongering. But a big, diverse family of renewable technologies could help bring down the UK's carbon footprint drastically.

Capturing the heat generated in power stations and using it to heat homes and factories could be a first major step. Replacing fossil-fuel-fired power stations with renewables can also make a major impact: every gigawatt of power replaced saves up to 1.5 million tonnes of carbon. If all the UK's power stations switched to renewable energy, that would save a huge proportion of our carbon footprint.

renewable energy

Renewable energy comes from sources that don't run out – like the sun, wind, tides, waves or plants. These natural sources can be harnessed to create electricity without adding carbon dioxide to the air.

The UK has signed a Europe-wide agreement to get 20 per cent of all energy (not just electricity but heat and transport fuel too) from renewables by 2020. We have a long way to go to reach this benchmark - in 2006 just 2 per cent of all UK energy came from renewables.

how much would it cost?

In any argument over the potential for renewable energy in the UK, the subject of cost inevitably comes up.

Because many of the technologies for producing renewable energy are relatively new their costs can be high, although wind turbines, for example, have become cheaper as the technology has matured. As confidence increases, however, the risk for investors gets lower, and attracting money for new projects becomes easier.

Costs can also be affected by other factors such as location – windy hilltops are better sites for wind farms than are sheltered valleys, and some stretches of coast offer greater wave potential than others. This means that the best sites – and the ones likely to be chosen first by developers – may produce more energy than ones set up later, even though the latter may have better technology. Working out costs also relies on a number of estimates, such as how much building and maintenance will cost in years to come. Ironically, one thing that historically has been left out of conventional economic assessments of energy is the cost of pollution.

The government could make cost a less significant barrier to renewables really taking off. It could, for example, remove some of the hurdles in developing wind farms, provide more subsidies for renewables and make it more expensive to pollute (see Chapter 6 for more on the government's role). Indeed, some argue that when it comes to tackling climate change the government should be doing everything in its power to make the shift to clean energy very rapidly.

what determines cost?

- Price of the raw materials for building the power station, wind turbine, solar panel
- Price of the land where the power station is built
- Price of building/installing the power station
- Price of connecting it to a local or national electricity grid
- Price of fuel
- Price of labour required to run the power station
- Amount of subsidy received
- Amount of electricity generated – and the market price
- Price of cleaning up pollution from the site
- Price of dismantling the power station at the end of its life.

load factor

Few sources of power provide electricity every day around the clock. Coal-fired, gas-fired and nuclear power stations need to be shut down for maintenance; wind farms, wave farms and solar installations work only when the wind blows, the tide flows or the sun shines. The level of output that any one source really works at is described as the 'load factor' and is expressed as a percentage of what an installation's output would be if it was working around the clock. For example, a power station that worked just three days in four over the course of a year would have a load factor of 75 per cent. In fact, conventional power stations had a load factor of 50 per cent in 2006.

RENEWABLE SOURCES OF ELECTRICITY

What are the alternatives to fossil fuels? Can we really run our homes, hospitals and holidays on the power of the sun and wind, waves and crops?

WEATHER PROFITS:

North Hoyle off shore wind farm, aproximately 10 kilometers off Rhyl in North Wales, is one of 140 wind farms in the UK supplying around 1 million homes with electricity.

wind power

Wind power turbines can work on land (onshore) or in the sea (offshore). The basic mechanism is the same for both but the practicalities, and therefore the potential, are different. Although under-exploited, wind power is well understood and already working hard for the UK. It is the most technically and economically developed of the renewables here – Britain's first wind farm, on Orkney, dates back to 1978, and the first commercial installation started up in 1991 at Delabole in Cornwall, where today ten turbines provide enough energy to power more than 2,000 homes.

Most wind turbines have two or three blades that rotate. The head of the turbine swivels to face into the wind, which drives a shaft to power a generator. Electricity travels via cables to a transformer, where it is converted to high-voltage electricity and fed into the national grid. Wind speeds need to reach around 10 mph before the turbines start generating electricity.

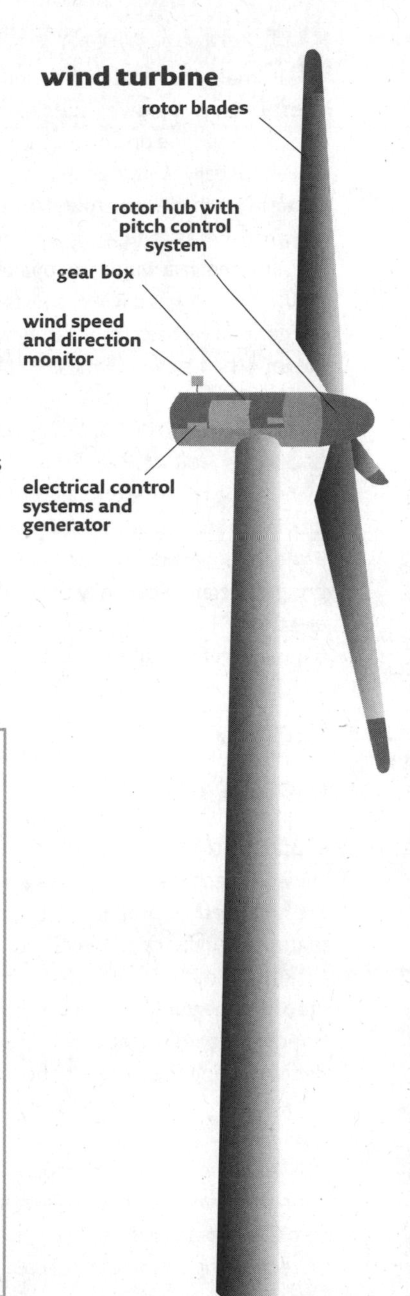

onshore wind power

Onshore wind at a glance

Fuel	Wind
Typical capacity	2-3 MW per turbine
Load factor	26-31 per cent
Land requirements	Small
Lifespan	20 years per turbine
Decommissioning	No lasting impacts
Pros	Zero net emissions after 3-5 months; inexhaustible resource; tends to be most productive in winter when demand is high.
Cons	Controversial visual effects; potential noise; interference with telecommunications; small threat to birds; variable output.

To make the most of the wind, turbines are put on exposed sites. They have grown in size as the technology has developed: a modern turbine on land has a capacity of around 2 MW, with several usually built together. A 5 MW wind farm can provide enough electricity for around 3,000 homes, depending on the site.

On average a wind farm operates at around 30 per cent load factor, but at a good windy site, this may increase to around 40 per cent. In very stormy weather wind farms shut down to prevent them being damaged.

what potential does onshore wind power have?

By mid 2007 there were 140 onshore wind farms in the UK providing enough energy for more than 1 million homes. Projects under construction or with planning permission would bring generating capacity to more than 3 per cent of the UK's electricity. But onshore wind projects – enough to generate 6 per cent of the UK's electricity – were, in mid 2007, still stuck in the planning system.

So what's stopping this well-understood technology coming into its own? A big obstacle is the way local council planning committees and planning officers are applying national guidance. Although planning applications are supposed to be processed within 16 weeks, for wind farms they are each taking an average of three years, and very few are getting the go-ahead. The government has amended planning guidance to make it easier for councils to prioritise tackling climate change, but many local councillors are ignoring the guidance. According to Alison Hill of the British Wind Energy Association, 'There's inconsistency in the decisions and then there's the time being taken. Councils are refusing wind farms for reasons that contravene planning guidance.'

Aside from planning, another obstacle is getting a connection to the National Grid. In 2006 projects totalling some 13.5 GW of potential wind power (including on- and offshore) were queuing up to be connected. Some do not expect to get a connection before 2015. Small-scale developments (under 5 MW for northern Scotland) can avoid this problem by being part of a local distribution grid. Wind farm developers pay for their connection to the grid, and in remote areas this pushes up costs. And each wind farm has to bid for a connection – a process that some say the industry regulator could speed up.

Some projects have run into delays because of increasing costs and waiting lists for turbines. But as the industry matures, new manufacturing plants should deal with this problem.

Some critics argue that wind power is of limited value because of the variability of the wind. But as one source of power contributing to a grid, this should not be a problem. National Grid says that we could cope with 20 per cent of electricity from wind before needing any extra back-up.

The government estimates that the practical potential for wind in the UK is 50 TWh – about 15 per cent of today's consumption of electricity. Others, including people in the industry, put the potential higher and say the future is for small wind farms of up to six turbines generating power for every town. With other efficient sources and renewables, wind power could be part of a mixed and decentralised local energy supply, linked on a low-voltage local grid.

offshore wind power

Offshore turbines stand on foundations driven into the sea bed with the head above the surface and swivelling to face into the wind. Newer offshore wind farms will be using larger turbines, producing more electricity. Innovative designs such as the Aerogenerator – which rotates on a vertical axis making it more robust in offshore conditions – promise 9 MW per turbine compared to the current 3 MW.

Offshore wind farms are more expensive to build and maintain than those onshore. The turbines need stronger foundations, and have to be protected against rusting from salt water. Connecting to the grid can cost more than onshore. But people tend to be less bothered by the visual effects of offshore installations – and offshore can generate a lot more power.

Offshore wind at a glance

Fuel	Wind
Typical capacity	3 MW per turbine; 5 MW turbines are being introduced
Load factor	35+ per cent
Land requirements	Offshore
Lifespan	20 years
Decommissioning	No lasting impacts
Pros	Zero emissions after less than a year. Visual effects less controversial than onshore wind, larger turbines can be used and winds are stronger. Can have positive effect on marine environment. Practical potential to supply quarter of UK electricity.
Cons	Possible effects on sea birds and the marine environment. Variable output. Needs new grid.

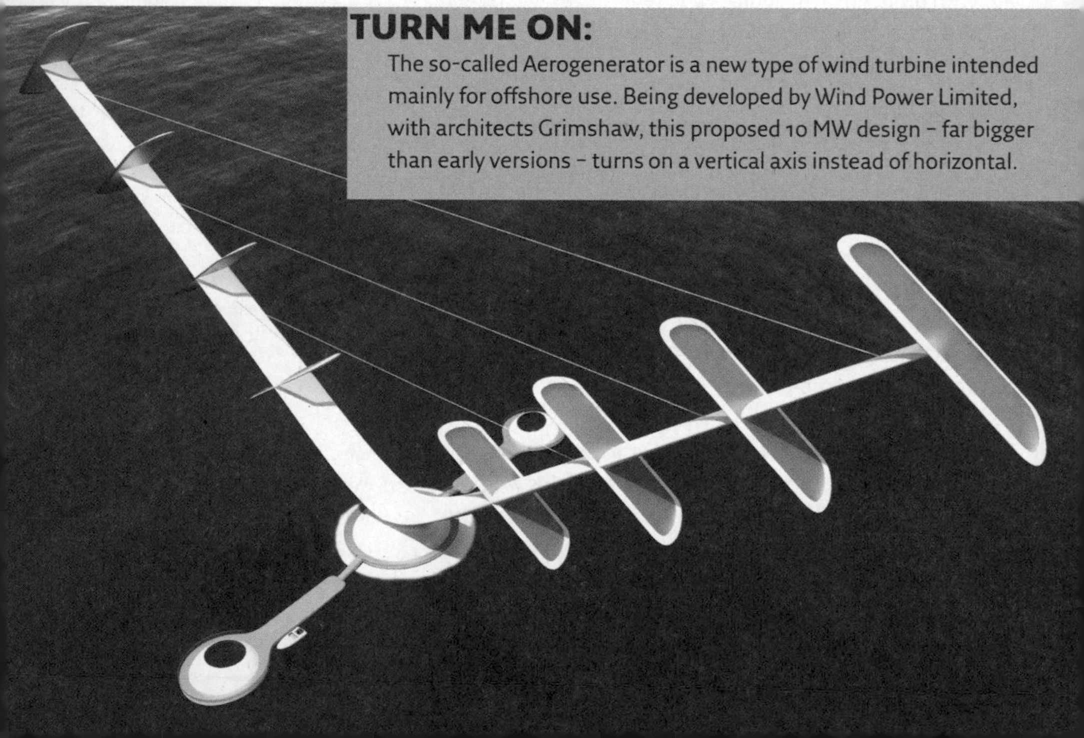

TURN ME ON:

The so-called Aerogenerator is a new type of wind turbine intended mainly for offshore use. Being developed by Wind Power Limited, with architects Grimshaw, this proposed 10 MW design – far bigger than early versions – turns on a vertical axis instead of horizontal.

five myths about wind turbines

1 "Wind power is unreliable and needs back-up from other sources"

Fact

Wind can be accurately predicted, giving a clear guide to how much power is likely to be available. National Grid, which is responsible for maintaining Britain's electricity supply, says wind power does not pose a major problem in balancing the UK's electricity supply.

2 "Wind turbines are dangerous for birds"

Fact

According to the RSPB, appropriately sited wind farms do not pose a big hazard for birds. If poorly-sited they can cause problems, especially where they damage valuable habitat. The RSPB warns that climate change poses the most serious threat to birds and wildlife and it supports the development of wind energy.

3 "Wind turbines are a blot on the landscape"

Fact

Some people love them, some hate them. Public opinion surveys generally show around 80 per cent support for wind energy. Given the choice between a wind farm, a coal-fired power station or a nuclear plant near your home, which would you choose?

4 "Wind farms bring down house prices"

Fact

Studies show that wind farms do not have a long-term effect on local house prices; prices may fall during the planning stage when there is uncertainty about the development. Improvements have overcome initial problems with noise, provided the turbines are appropriately located.

5 "It takes more energy to build a wind farm than it will ever generate"

Fact

According to npower, the average wind farm will pay back the energy used in its manufacture within just three to five months of operation. Given an expected lifespan of 20 years, that's 19 years and seven months of carbon-free energy.

battle on the Marsh

The Little Cheyne Court Wind Farm, near Lydd in Kent, was expected to start producing electricity towards the end of 2008, six years after it was first proposed. When complete 26 wind turbines standing on Romney Marsh would supply around 30,000 homes.

The site was controversial. Romney Marsh, a vast windswept tract in the south-east corner of England, is important for birds. Large areas lie below sea level and salt-water lakes are dotted across the landscape. Neighbouring Dungeness, home to one of Britain's nuclear power stations, is also the world's biggest expanse of shingle, a National Nature Reserve and internationally important wildlife site. The RSPB reserve attracts a rich variety of birds including smew, wheatears and bitterns.

When Shepway Friends of the Earth local campaigner Barrie Botley first heard about the wind farm proposals, he says his reaction was mixed. 'We really did not know enough about the bird life,' he recalls. 'But being so close to the nuclear power station at Dungeness, I did think it would be one in the eye to get a wind farm near by.' The Friends of the Earth group contacted the RSPB and discovered that visiting birds appeared to have learnt to avoid the power cables spreading out across the Marsh from Dungeness power station.

So the group decided to back the scheme. Some locals were less convinced and a row ensued in the letters pages of the local press. 'We had a letter a week for over a year,' Barrie remembers. 'We wanted to reach out to a wide spectrum of people and persuade them that the wind farm was a good thing.'

The group did its research, drawing on studies of wind farms in the Netherlands to challenge the idea that birds would be harmed. Group members went to public meetings and set up a rapid response unit to deal with opposition in the local press.

'People in the area are really in love with the countryside, but they exaggerated claims as to how it was going to devastate the whole of the Marsh,' Barrie says. 'It got quite nasty at times.'

Local councillors opposed planning permission for the site. But at a public inquiry the government gave the go-ahead for 26 turbines. When up and running, Little Cheyne Court will be the most powerful onshore wind farm in the south of England.

what potential does offshore wind power have?

With shallow waters and strong winds, the UK has potentially the largest offshore wind resource in the world. Some estimates put this at a third of the total offshore potential for Europe – enough electricity to power the UK several times over. A government study said that by 2017 offshore wind could provide the equivalent of about a fifth of today's total consumption.

Yet the UK has been slow to get off the mark. The Horns Rev wind farm, off the coast of Denmark, has for some time been the world's largest with 80 turbines. The UK had five working offshore wind farms in 2007; six under construction would still bring the total to less than 1 GW of installed capacity. A big new wind farm, the so-called London Array, however, is due to start up in 2012 with around 300 turbines and a total capacity of 1 GW (enough for 750,000 homes). Seven larger sites that already have permission, plus applications for more, could bring capacity up to around 7 per cent of the UK's electricity supply.

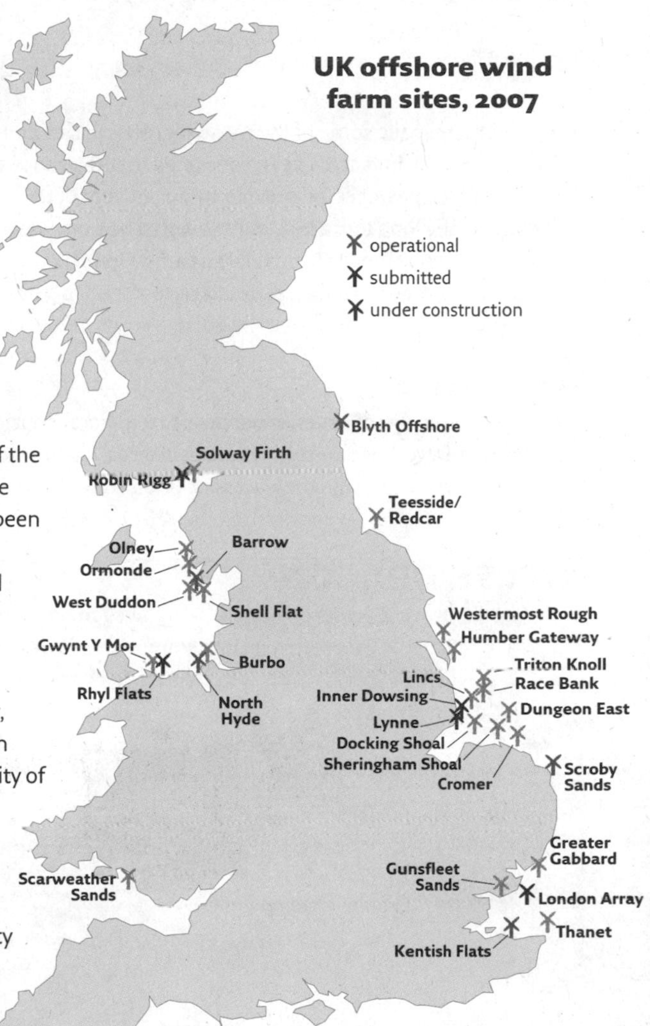

power from the sea

The British Isles have some of the most energetic waves and tides in the world. In fact it's been estimated that the various forms of marine power could provide 15-20 per cent of UK electricity in the long term. Most of this would be from wave energy with a tiny amount from tidal streams. One study estimated that marine resources could generate up to 3 per cent of the UK's electricity supply by 2020.

wave power

We can tap the energy of waves, either close to the shore or out in deeper water. Developers are testing various designs, from cylinders that move up and down with the waves out at sea, to systems

WELSH DRAGON:
A prototype wave power device, the Wave Dragon (see p.146) is being tested off the Pembrokeshire coast, taking the first step in establishing a 70 MW wave power plant in the Celtic Sea.

attached to the shore or built in to harbour walls. The Limpet traps incoming waves, creating a moving column of water that can be used to generate power. The Archimedes Wave Swing (AWS), a cylinder-shaped buoy attached to the sea bed, is filled with gas that is compressed by the pressure of the waves. A prototype is being tested off the coast of Orkney. AWS Ocean Energy, due to develop a commercial site in 2009, says the technology could deliver 100 MW of electricity from just 1 square kilometre of seabed.

Wave power at a glance

Fuel	Waves
Load factor	30 per cent
Typical capacity	3-7 MW (for an installation of four units)
Area requirements	Offshore 1 square km could provide 80 GWh per year
Lifespan	Approx 25 years (Pelamis)
Pros	Zero carbon emissions after 20 months. Low visual and few environmental effects.
Cons	Cost. Still at the pilot stage. Access to the grid problematic in areas with the most potential. Possible impacts on shipping, fishing and marine life.

what potential does wave power have?

No wave power technology has yet been put through its paces on a commercial scale. Although studies point to the UK having some of the best marine and tidal stream potential in Europe, the best sites for wave technology are off the west coast of Scotland, where there is little carrying capacity in the grid.

Developing new technology offshore is expensive, but wave power is expected to get cheaper as the technologies advance

snakes and dragons

The Snake: An early front-runner among wave power devices, the Pelamis or Sea Snake is made up of linked cylinders partly submerged. As the cylinders move with the waves, oil is pumped under pressure into motors which drive generators creating electricity. In a pilot project four cylinders, about the length of five train carriages, are linked together, with the potential for several links to line up in parallel to form a wave farm. The first commercial plant is being installed off the Portuguese coast. Scottish Power was due to start testing the Pelamis off Orkney in 2008 with a single strand of four cylinders, expected to generate enough electricity for 3,000 homes.

The Dragon: The first and largest UK offshore wave energy installation, the Wave Dragon was commissioned in 2007 off Milford Haven. A moored device with the capacity to power 2,000-4,000 homes, the Dragon acts like a funnel to channel waves into a reservoir above sea level. The water is then released through turbines to produce electricity. The company behind the Wave Dragon plans a 70 MW plant in the Celtic Sea by 2010.

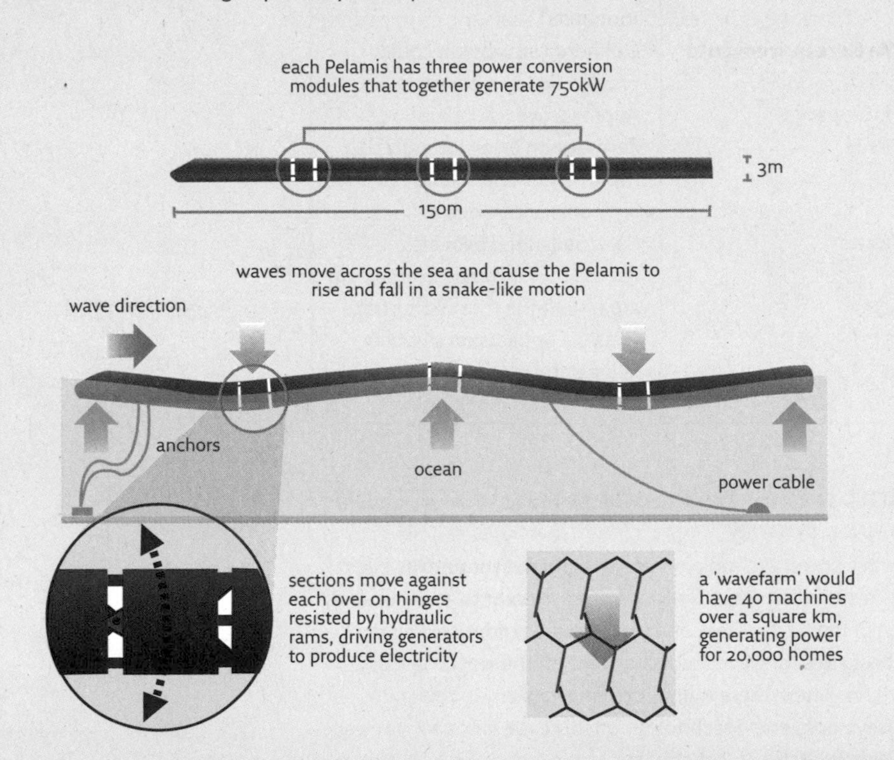

and to become commercially viable if financial support is made available soon. The UK is among the leading countries developing marine energy technologies. Being at the forefront offers potential in terms of jobs and revenue. According to the Carbon Trust, worldwide electricity revenues from wave and tidal projects could reach £60–£190 billion a year.

What will wave power do to the environment? Because the devices have so far been tested on a small scale, it is difficult to say. Although installation is likely to disrupt the seabed, in the long term they could provide new habitat for sea life.

tidal power

The power of the ebb and flow of the tides can be captured in tidal estuaries. Any stretch of water where the difference between low and high tide is more than 8 metres – such as the Irish Sea, the outer Bristol Channel, the Wash, the eastern end of the English Channel and the Channel Islands – is considered to have the potential to develop a tidal energy plant.

Different ways of capturing this energy are being evaluated. One approach is to use underwater turbines, another is to bridge estuaries (see Tidal barrage, below) and yet another creates lagoons that retain water as the tide goes out and then release it through turbines (see Tidal lagoons, below).

what potential does tidal power have?

Tidal power offers predictable energy but its future contribution will be in part limited by the number of suitable locations around Britain's coasts. The best potential is in the Pentland Firth in north-east Scotland but the remote location is a problem.

One study has suggested that the top ten tidal sites could produce around a tenth of our electricity. Business lobby group Scottish Enterprise suggests that 34 per cent of total demand is possible. Government predictions are much less ambitious, saying 0.27 per cent of our electricity could come from tidal power by 2017. But tidal power developer Martin Wright says the government's approach lacks vision. 'I don't believe they understand what renewable energy can do,' he says. 'We need to be thinking more radically.'

SEA CHANGE:

Before the days of cheap coal numerous coastal communities used the ebb and flow of tides to produce power for local industry and electricity. The time has come to revive the idea.

tidal barrage

A tidal barrage dams an estuary with turbines powered by the incoming or outgoing tides – or both. The turbines are built into a dam wall, which holds back the incoming water to maximise power.

The only existing tidal barrage is on the Rance Estuary in France. Built in 1967, it powers about 250,000 homes, but has caused silting, halving the estuary's tidal range and changing marine life.

A much bigger, 10-mile tidal barrage across the Severn Estuary has been proposed. This would raise the water level in parts of the estuary by up to 5 metres and flood around 70 per cent of the inter-tidal area. The scheme has drawn fierce criticism for a number of reasons – from its high cost (compared to tidal lagoons, for example) to the fact that the estuary is an important feeding ground for up to 65,000 migrating birds and provides unique wildlife habitats.

Tidal barrage at a glance	
Fuel	Tides
Load factor	Around 25 per cent
Land requirements	Large infrastructure in tidal estuaries (Severn barrage would impound 185 sq miles)
Lifespan	120 years +
Decommissioning	Could be maintained as a road and rail link
Pros	Relatively large output for one installation; flood protection; can benefit some species.
Cons	Intermittent; very limited capacity to store energy; cost; reduces tidal reach in the estuary, impacts on birds and marine life; high visual effect; can affect shipping.

what potential do tidal barrages have?

The Severn has the second highest tidal range of any estuary in the world and the proposed Severn barrage is predicted to generate just over 4 per cent of UK electricity. The idea of a barrage on the site has been around since the 1970s but high costs, the effect on ports in the Estuary, and fears for wildlife have prevented it from going ahead. In 2007 the government's sustainability watchdog said feasibility studies should begin on the barrage, and dismissed alternative proposals for large lagoons.

Electricity generation from a tidal barrage, while predictable, is highly intermittent. The Severn barrage would produce two large surges of power for a few hours twice a day, and these would only occasionally coincide with peak demand.

tidal lagoons

Like barrages, lagoons are artificial enclosures offshore in areas where there's a big difference between high and low tide. As the sea rises and falls water enters or leaves the lagoon, passing through turbines in its walls. Lagoons can also be used to store and release water, creating a potential reserve electricity supply. Tidal lagoons can be large or small and are able to generate power on both incoming and outgoing tides.

Lagoons at a glance

Fuel	Tides
Load factor	33–57 per cent, depending on location
Land requirements	Infrastructure located in shallow coastal waters, can be tens of square miles
Lifespan	120 years +
Pros	Does not impound sensitive inter-tidal areas; does not impede shipping; relatively low visual effect; creates new marine habitat; potential to store and supply energy on demand.
Cons	Not tested commercially; may cause silting; some disruption to traffic in the estuary

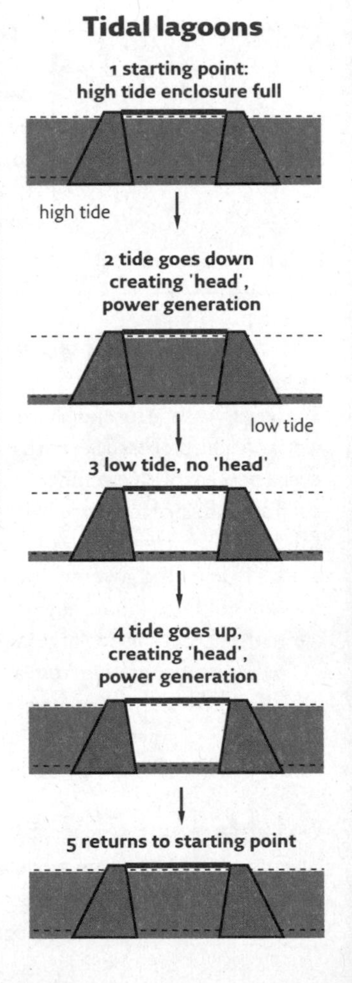

what potential do tidal lagoons have?

Tidal Electric, the company behind tidal lagoons in the UK, claims that a lagoon enclosing 50 square miles of the Severn Estuary across the tidal range could spread generating capacity over longer periods of time than the proposed barrage, creating less of a surge in power and so avoiding one of the barrage's problems. The company says lagoons in the Severn Estuary could supply up to a quarter of today's UK demand. The actual level of output would depend on environmental considerations and the viability of the technology. Lack of government support has discouraged developers from applying for consent to build a lagoon.

underwater windmills

The power of the tides is being tested in Strangford Lough in Northern Ireland. A SeaGen turbine was set to be installed underwater here in 2008, generating enough power for around 1,100 homes. The twin-rotor turbines sit below sea level, attached to a steel tube driven into the seabed. The in-coming and out-going tides turn the blades. Because they rotate slowly – at around half the speed of most ship propellers – they probably pose little threat to marine life. Strangford Lough is an important wildlife site and the turbine will be monitored to ensure it does no damage.

Martin Wright of Marine Current Turbines, the company behind SeaGen, explains that Strangford Lough is an ideal site to test the turbine as it has shallow water and easy access. He acknowledges, however, that the turbine will have to show that it does not damage the unique habitat in the Lough. 'We will stand or fall by whether this is as environmentally benign as we think it is,' he says. If the pilot succeeds, the company hopes to start work on a bigger project off North Wales in 2009.

power from the sun

Electricity from the sun's rays comes in two forms, solar photo voltaic (PV) power, and concentrated solar power (CSP).

Solar photo voltaic (PV) cells are designed mainly for small-scale or micro-generation. They are made from semi-conducting materials, such as silicon, that produce electricity when exposed to the sun. Fitted into panels, PV cells can be used on roofs, as cladding on buildings, or as standalone or even portable generators. The charge from the PV cells is converted to alternating current for home, office or any other building. Excess electricity can be fed into the grid.

solar PV

Cells vary in design. Some types contain cadmium, which is toxic. Although the technology has a high initial cost, if integrated during building or refurbishment it can replace roof tiles or cladding. Once installed it generally requires little or no maintenance. New cells are being tested that can work indoors and in dim conditions.

Solar PV at a glance

Fuel	Sunlight
Output	0.1 MW per square metre
Land requirements	Can be fitted on existing roofs and buildings
Lifespan	30 years +
Pros	Inexhaustible; lots of roof space available; offgrid potential
Cons	Cost; energy intensive in the manufacturing stage

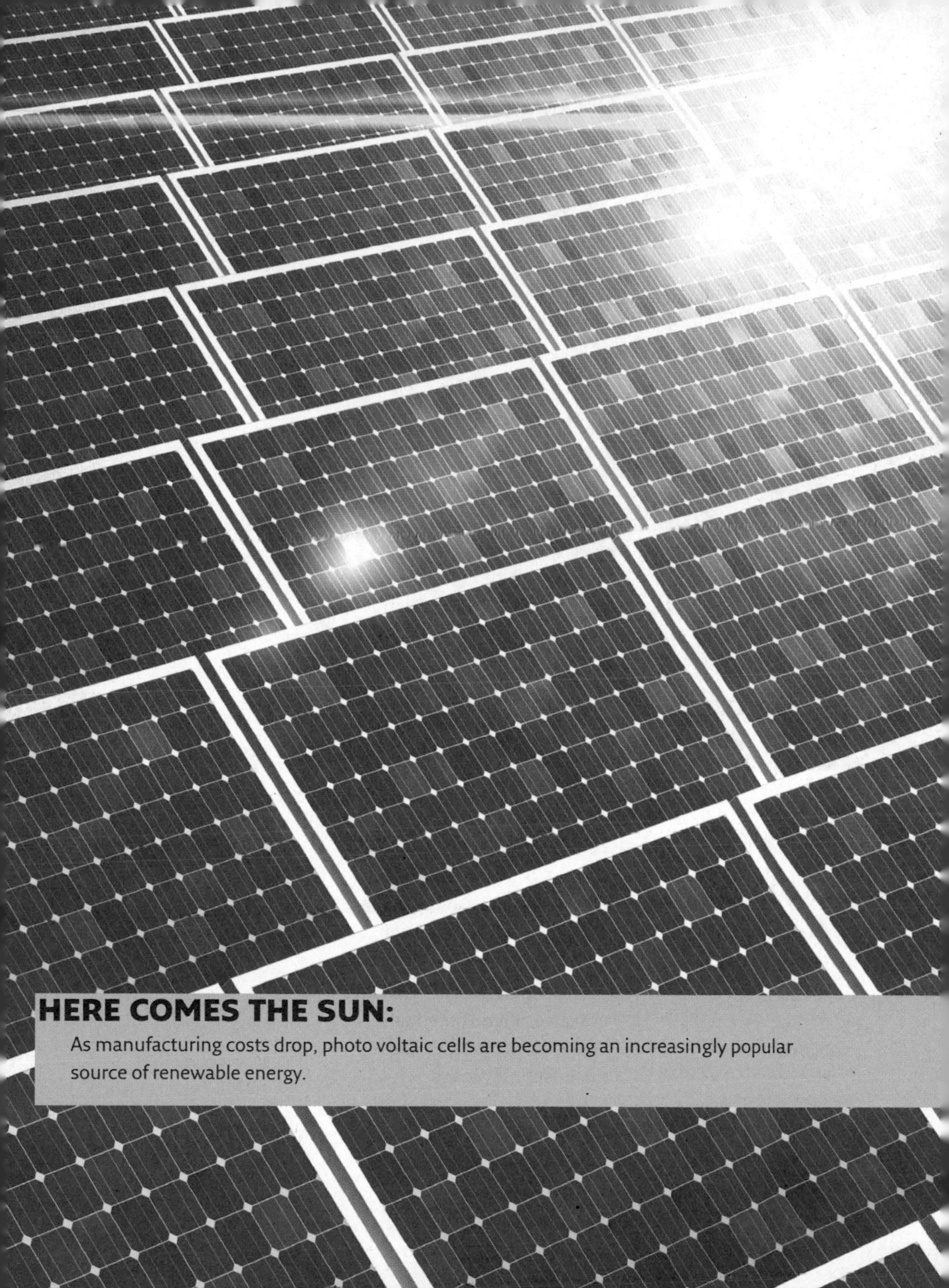

HERE COMES THE SUN:
As manufacturing costs drop, photo voltaic cells are becoming an increasingly popular source of renewable energy.

BANKING ON THE SUN: The Cooperative Insurance Group clad the outside of its 40-year old grade II listed tower in the centre of Manchester with 7,000 photovoltaic panels that produce enough electricity to power 55 homes for a year.

what potential does solar PV power have?

PV cells are most effective in a south-facing location (in the northern hemisphere) but don't need to be in direct sunlight to work – in fact they will produce electricity even on cloudy days (but not at night). Power varies according to the position, the amount of sunlight, and the type of cell used. Installer Solarcentury says 10 square metres of crystalline panels will provide nearly a third of the electricity needed by a three-bedroomed home.

The government sees the potential for solar PV as a tiny 0.03 per cent of total electricity demand. But if prices fall, it could become more prominent. Advocates of solar power argue that the government has not provided the support that would help the industry become competitive. According to Seb Berry of Solarcentury: 'There is clearly a view at the highest levels of government that "grown ups" don't get their energy from renewable, community or micro-generated sources.'

In Germany by comparison the government has boosted renewable energy via a feed-in tariff (see p. 227) guaranteeing a premium rate for renewables. Germany has 2 GW of solar PV – the equivalent of 800,000 domestic solar roofs – and manufacturers there say the technology is likely to become competitive within three to ten years. Although the UK has less sunshine than many of its European neighbours, solar power can still make a valuable contribution. Commercial buildings are beginning to take advantage of the technology. The new PV façade on the CIS Solar Tower in Manchester, for example, cost £5.5 million; but every year it saves more than 100 tonnes of carbon dioxide – the equivalent of 20 hot air balloons filled with the greenhouse gas.

concentrated solar power (CSP)

CSP at a glance

Fuel	Sun
Typical capacity	Up to 200 MW/ scheme
Land requirements	Vast uninhabited desert areas available
Pros	Enormous potential; carbon-free, safe; probably low ecological impact; potential major resource in sunbelt regions; large storage capability; could provide shade for agriculture and homes.
Cons	Not suitable for UK weather but could be connected via European grid; visual effects.

Concentrated solar power uses mirrors to focus the sun's heat onto a steam generator and produce electricity. In a trough system mirrors reflect sunlight onto pipes containing oil, which transport the heat to a conventional steam generator.

Europe's first commercial CSP generator, the Solair in southern Spain uses 600 mirrors that track the sun as it moves and focus it onto a receiver on a 115 metre-high tower. The receiver transfers the heat, via a column of hot air, to drive a steam generator and produce electricity. The Spanish plant produces enough to power 6,000 homes – and as more fields of mirrors are added, could power the nearby city of Seville.

MIRROR, MIRROR:

By concentrating the enormous power of the sun, CSP can provide clean electricity. The technology is not well-known in countries which aren't very sunny (such as the UK) but it has the potential to vastly reduce the world's consumption of fossil fuels.

Early CSP relied on the additional burning of natural gas to keep a steady basic temperature. But newer systems require minimal back-up. The Nevada Solar One, in the desert near Las Vegas, needs just 2 per cent gas back-up to keep it going.

what potential does CSP have?

Concentrated solar power could make a big contribution where there is a lot of sunshine. Desert areas such as the Sahara, southern Spain and California are ideal.

CSP is flexible – it has the potential to generate electricity for the grid or to work at a local level powering a factory or village. Heat from concentrated solar power can be stored in water tanks for use at night or on cloudy days. An additional advantage is that the sea water used for the steam process can be desalinated to provide plentiful supplies of potable water.

Some studies suggest CSP sites in Southern Europe and North Africa could generate enough to replace all of Europe's nuclear power and vastly reduce the consumption of fossil fuels. Covering just 1 per cent of the world's deserts with CSP could produce enough electricity to meet global demand. The key drawback is that investment would be needed in high-voltage cables linking the deserts to the areas using the power. Even so, it's thought that the UK could be importing small quantities of CSP by 2020. Algeria is aiming to export 6 GW of CSP to Europe by that date.

hydro-electric power

Hydro-electric power is derived from lakes, reservoirs, rivers and streams.

small-scale hydro-electric power

Water power has been used for centuries – from corn mills to the cotton mills of the Industrial Revolution. More efficient turbines replaced water wheels in the 19th century and a few still work today. Small-scale hydro (anything producing 1-20 MW) can use flowing water to drive a turbine. The electricity can be used locally or fed into the grid.

Most small-scale hydro schemes divert water from a river or stream into the turbine. To have enough power to generate electricity the river or stream needs to have a drop height of at least 2 metres and a reasonable volume of water. This drop height can be created by a weir, where turbines can also be housed.

Where migrating salmon or trout use the river, the fish must be given an alternate route – sometimes a series of pools staggered up the slope.

Small-scale hydro at a glance	
Fuel	Flowing water
Typical capacity	1-20MW
Load factor	40 per cent
Lifespan	50 years +
Pros	Local; very low carbon; reliable; low-tech; off-grid potential
Cons	Reduced potential during droughts; some effect on ecology of water course; initial cost.

what potential does small-scale hydro-electric power have?

In 2007 the UK had slightly more installed capacity for small-scale hydro than for offshore wind. The greatest potential tends to be in mountainous areas like the Scottish Highlands where there are large waterfalls, but lower-lying rivers can be used if there is a strong enough flow. Experts estimate that turbines at the 44 existing weirs along the length of the River Thames could generate enough for up to 1,200 homes.

Small-scale projects – especially where there is only a low drop – can be expensive to install, but the technology is efficient: small-scale hydro converts between 50 and 90 per cent of the available energy into power. There is potential for small-scale hydro to meet 3 per cent of the UK's electricity demand.

the turn of the screw

In the 19th century the Derbyshire town of New Mills thrived, thanks to the rivers running through the Torrs gorge that powered the town's cotton mill. Today that power is again providing electricity for local people.

A social enterprise has formed a partnership with the community to develop a small-scale hydro scheme on the site of one of the town's old mills. This uses an Archimedean screw installed alongside a weir on the former mill site. Installed in 2008, the scheme is expected to generate 260 GWh a year – enough to power the local school.

Richard Body, from High Peak Friends of the Earth group, would like to see more hydro projects in the area. 'It's just common sense,' he says. 'It is a good practical way of generating electricity for practically no cost and with no environmental impact. I'd like to see hydro schemes on all the weirs on the river.'

Other sites have been identified in Wakefield and Sheffield. Weir by weir, the carbon savings could be adding up.

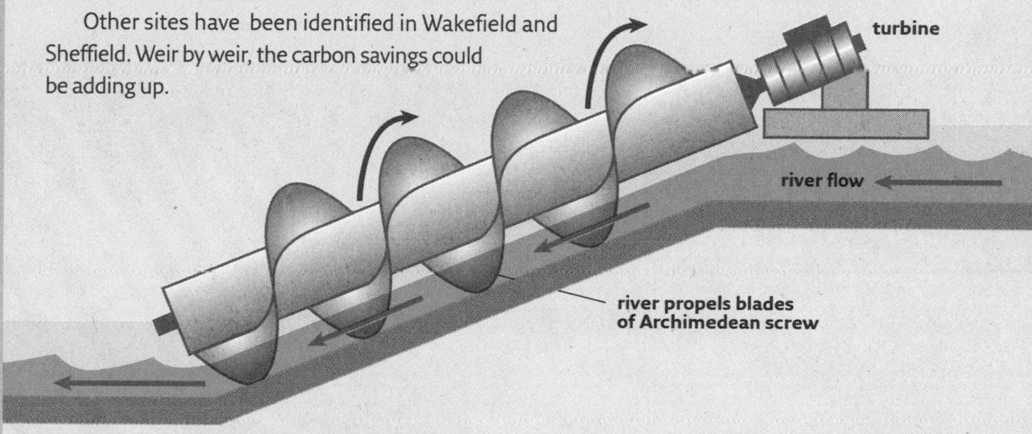

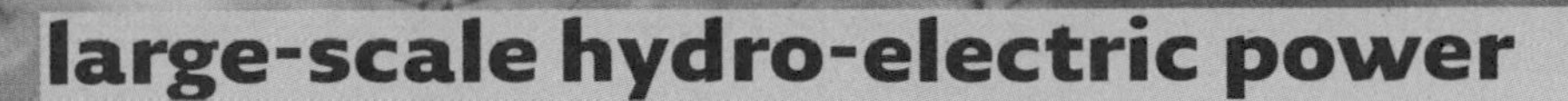

large-scale hydro-electric power

Large-scale hydro-electric dams and reservoirs generate a fifth of the world's electricity. The world's biggest is the Itaipu project shared by Brazil and Paraguay which has a capacity of 12.6 GW. The United States, Canada and China are the three largest generators of hydro-electricity. China's Three Gorges Dam will have a capacity of 22.5 GW.

Such huge dams can damage the local environment, changing the flow of the river, and flooding large areas. In some parts of the world, whole communities have been wiped off the map by the reservoirs. The UK's large hydro projects, built in the 1950s, do not qualify for the Renewables Obligation (see p.226) because they have already paid for their construction costs.

Hydro-electric power already provides a back-up system for the UK grid: at Dinorwig and Ffestiniog in Wales, water is pumped from a lower reservoir when excess power is available; and when more power is needed the water in the upper reservoir is released, rushing through turbines to generate instant electricity for the grid.

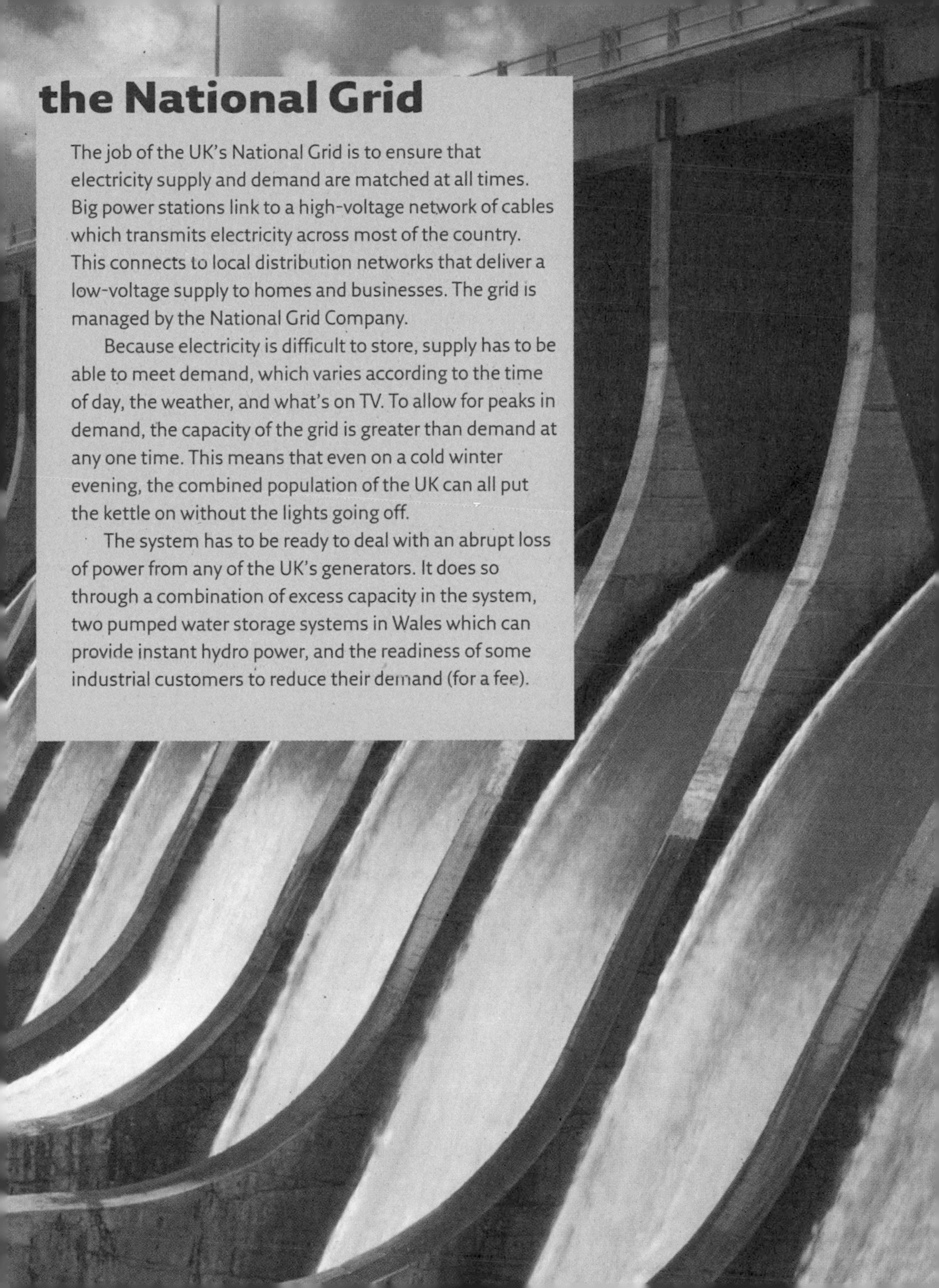

the National Grid

The job of the UK's National Grid is to ensure that electricity supply and demand are matched at all times. Big power stations link to a high-voltage network of cables which transmits electricity across most of the country. This connects to local distribution networks that deliver a low-voltage supply to homes and businesses. The grid is managed by the National Grid Company.

Because electricity is difficult to store, supply has to be able to meet demand, which varies according to the time of day, the weather, and what's on TV. To allow for peaks in demand, the capacity of the grid is greater than demand at any one time. This means that even on a cold winter evening, the combined population of the UK can all put the kettle on without the lights going off.

The system has to be ready to deal with an abrupt loss of power from any of the UK's generators. It does so through a combination of excess capacity in the system, two pumped water storage systems in Wales which can provide instant hydro power, and the readiness of some industrial customers to reduce their demand (for a fee).

biomass – energy from organic matter

We can turn trees, plants and other organic material such as food scraps into energy for heat and electricity. This is sometimes referred to as biomass, bio-energy or biofuel. Although waste products like wood offcuts or wood chips can be used, high demand for biomass would lead to farming of energy crops on industrial-scale monoculture plantations.

Biomass at a glance

Fuel	Energy crops, wood and food waste, slurry
Load factor	60-85 per cent
Land requirements	Farmland to grow crops, and woodland
Pros	Renewable, productive use for waste, can be locally sourced
Cons	Large areas of land to grow significant amounts; social and environmental impacts of big plantations; short supply of local biomass; competes with food; potentially emits more greenhouse gases than fossil fuels.

In power stations biomass can replace a proportion of the coal or gas. It could be a carbon-neutral energy source since the plants it comes from absorb carbon dioxide as they grow. But because of the huge quantities of energy and fertiliser that it takes to cultivate, process and transport many biomass crops, emissions can compare badly with fossil fuels.

FLAME TREES:

Biomass is probably the oldest form of fuel – think wood and cow dung – and it can play an important part in our overall energy mix. But industrial-scale energy crops come with a health warning for the environment.

what potential does biomass have?

The amount of plants and other organic material available will determine the potential for biomass. We produce lots of organic waste, and there is a certain amount of natural waste wood lying around. But a big shift to this method of producing energy would imply industrial farming and importing energy crops, which can pose serious social and environmental problems. The UK produced 1.5 per cent of its electricity from biomass in 2003. The Aberthaw power station in Wales,
for example, burns 20 tonnes of wood chips per hour taken from sawmills and other waste wood, alongside its main source of fuel, coal. It is said that new types of biofuels could increase the amount that the UK could use in future but there are unanswered questions about the commercial viability, environmental impacts and carbon savings of second-generation fuels.

Friends of the Earth believes bio-energy has a role to play in bringing down greenhouse gas emissions only if it is done in a way that protects wildlife and people's livelihoods, and guarantees emissions cuts. Realistically this role will be a very small part of the overall energy mix.

adapting non-renewable sources

The more renewable power we develop, the more we reduce our carbon footprint – especially if the most polluting fuels like coal are replaced. The UK could in the future get all its electricity from clean renewable sources; but in the short term they won't meet all of our needs. And we need cuts urgently. So what's the future for non-renewable sources? Can we clean up existing fossil fuels? And what about nuclear power?

combined heat and power (CHP)

CHP at a glance

Fuel	Coal, gas, oil, biomass, nuclear
Load factor	70-85 per cent
Land requirements	Minimal (unless using biomass)
Pros	Very efficient use of fuel (e.g. could double efficiency of coal-fired power station); pipeline network would be long lasting.
Cons	Needs costly hot water pipeline network; limited biomass available – most generation from gas; large CHP more expensive than simple coal and gas generation; usually requires digging up roads or pavements.

CHP use in selected European countries

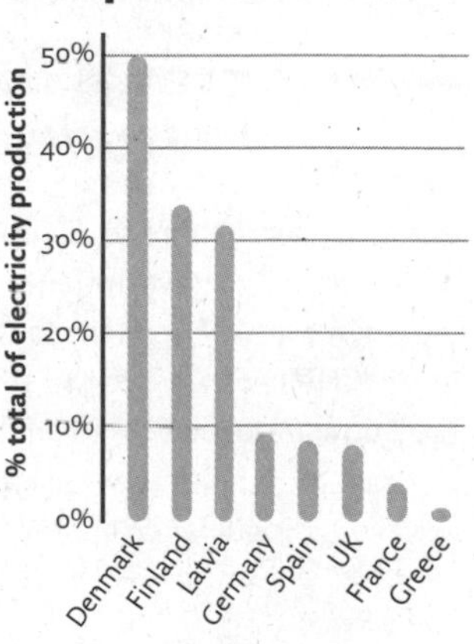

Source: European Environment Agency

Today we get electricity mostly from power stations and heat mostly from boilers. Simply producing heat and electricity from the same fuel in the same plant can increase efficiency massively, so shrinking the overall carbon footprint. This is combined heat and power (CHP) - also known as cogeneration.

Most CHP plants burn gas, although coal and biomass can also be used. In big power stations heat can be captured from the generation process and channelled to factories or nearby homes.

Smaller plants can supply heat and electricity to a housing estate or district, independent of a national grid – although in fact they are usually connected for security of supply. Odense in Denmark, for example, has a city-wide heating system that includes a coal-fired power station running at a claimed 90 per cent efficiency (more than double the efficiency of a non-CHP coal power station).

Domestic CHP boilers are being developed that can burn gas, oil or biomass such as wood pellets to produce heat, some of which is captured to generate electricity. This makes domestic CHP boilers far more efficient than conventional boilers.

what potential does CHP have?

Combined heat and power could reduce the UK's carbon footprint by making us much more energy efficient; but the UK currently lags behind its European neighbours in using this technology. The government has an aspirational target that would imply CHP providing a third of UK electricity demand by 2020. CHP can improve the efficiency of new large-scale gas-fired power stations by 30 per cent – and such plants are obliged to show that they have explored the option. Friends of the Earth believes this guidance should be strengthened.

woking wonders

Woking Borough Council has cut its carbon footprint by more than 80 per cent since 1990. Combined heat and power (CHP) has played a central role in achieving this.

The local authority developed one of the first CHP energy stations in the UK, which started generating in 2001, using gas to provide highly efficient heat, hot water, electricity and cool air for town centre buildings. Run by Thameswey Limited, a company set up by the council, the energy station has its own local private electricity grid, which is linked to the National Grid to guarantee a back-up supply. The project provides its own electricity more than 90 per cent of the time.

In 2007 there were some 14 CHP plants in the borough, powering sheltered housing schemes and a doctor's surgery. Some work alongside solar PV panels to provide heat, hot water and electricity and feed private electricity grids. At Woking's Pool in the Park, a gas-fired CHP plant works alongside a hydrogen fuel cell to heat the pool water and provide electricity. The town centre's energy station makes enough electricity for around 1,300 homes and heat for the equivalent of 500 households. Woking Borough Council's own offices use the power, as do residential and business customers.

A trailblazer in the energy sector, Woking Borough Council continues to introduce new technologies, with combined wind and solar PV-powered street lights and a solar PV gateway, providing a fitting entrance to the town.

carbon capture and storage (CCS)

CCS at a glance

Fuel	Coal, gas and/or biomass.
Land requirements	Sub-sea or other natural geological formations or disused oil and gas wells.
Lifespan	Variable depending on location.
Pros	Potential to quickly reduce emissions from existing plants.
Cons	Untested on commercial scale, potentially expensive, risk of carbon dioxide leaks.

How carbon capture and storage works

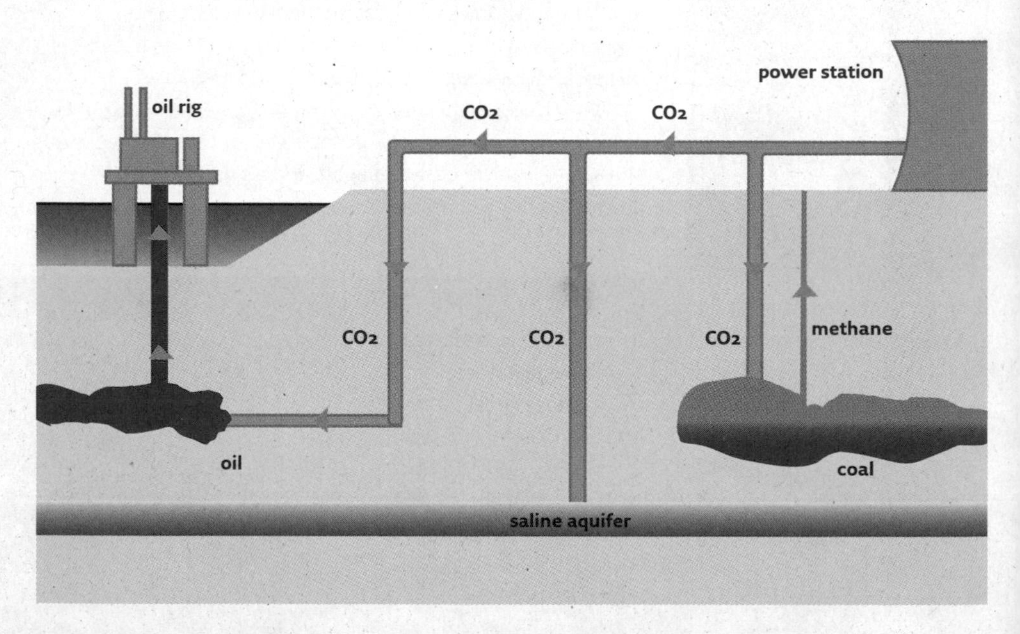

Carbon dioxide from power stations can be captured and the gas stored by pumping it into underground caverns. Empty or near-empty oil and gas fields in the North Sea have been proposed as one storage option, alongside saline aquifers (pockets of salty water underground) and coal mines that can no longer be mined.

The technology is already in use: in the United States and Canada carbon dioxide is pumped into depleted oil fields to help extract the remaining oil. But this so-called enhanced oil recovery has not been considered commercially viable for the UK's North Sea oil and gas fields. In Norway the oil company Statoil started pumping carbon dioxide into porous rock in 1996, storing around 1 million tonnes each year, as part of enhanced oil recovery. Oil company BP has plans for a scheme in Western Australia using gasified coal to make hydrogen for electricity, with 90 per cent of the carbon dioxide captured and stored underground. On Teesside, Progressive Energy is planning to use CCS with a clean-coal and CHP development – with the scheme potentially up and running by 2012.

There are concerns that underground storage might not be leak proof. While oil and gas fields trapped oil and gas for millions of years, drilling may have caused some damage. Proposals to store carbon dioxide on the ocean floor are thought to be far more risky as little is known about how this will affect marine ecosystems or the ocean's ability to continue to absorb carbon dioxide.

what potential does carbon capture and storage have?

CCS is seen as potentially making an important contribution to reducing carbon emissions by 2030. But the technology is not yet proven and regulating access to the seabed must be resolved. The UK Government aims to have a demonstration project operating by 2014.

The process of capturing carbon uses a lot of energy and will require special equipment at coal-fired power stations. Once captured, the carbon dioxide has to be transported, again adding to the cost.

Nevertheless CCS could provide a stop-gap way of shrinking our carbon footprint in the medium term, if incentives justified the investment. This is considered particularly important for countries such as China which have large supplies of coal. But the extra costs mean it may only be affordable when carbon reaches a high enough price.

Carbon capture could also be used at power stations burning carbon-neutral fuels such as biomass. This would have the net effect of reducing the amount of carbon dioxide in the atmosphere – creating a negative carbon footprint.

combined cycle gas turbines (CCGT)

CCGT at a glance

Fuel	Gas
Land requirements	Small
Lifespan	20+ years
Pros	Relatively clean, efficient fossil fuel or biofuel
Cons	Uses a fossil fuel so still emits carbon dioxide. Potential methane leaks. Limited biofuels, finite fossil fuels.

Combined cycle gas turbines (CCGT) are already used in some gas-fired power stations in the UK. They are more efficient, and therefore produce less carbon dioxide emissions, than conventional turbines as they use the heat produced by a standard gas turbine to make steam which drives a generator. The result is that around half the available energy is converted into electricity. The technique can be used in combined heat and power plants (see above), with the steam used to provide district heating. CCGT plants produce nitrogen oxide and carbon monoxide, but these can be reduced by using catalysts.

what potential does CCGT have?

Almost all gas-fired power stations built since the early 1990s have used CCGT technology. E.ON runs a CCGT plant at Connah's Quay on the Dee Estuary, providing enough electricity to power half of Wales.

Because gas has been cheap, the electricity generating industry sees CCGT plants as the power station of choice. Combining this technology with CHP – and building power plants close to towns so that the heat can be used locally – could make gas-fired power stations even more efficient.

new or 'clean' coal

New coal at a glance

Fuel	Coal (potential for some biomass)
Land requirements	Moderate – possible to upgrade existing plant
Lifespan	20+ years
Pros	Increased efficiency, potential for co-firing using carbon neutral fuel source, uses a lot of the existing plant and infrastructure; mature technology; existing plant is reaching the end of its natural life and needs replacing.
Cons	Relies on fossil fuel, produces lots of carbon dioxide; other impacts of the coal mining industry

There are more efficient ways of burning coal than is currently practiced in most UK power stations, which are 30 to 40 years old. But even the newer methods still only achieve 45 per cent

SHAFTED:
Whether deep-shaft or open cast, coal mining accesses vast quantities of carbon which release three times their weight in carbon dioxide when used in power stations.

efficiency. Until we start capturing waste heat as standard practice, 'clean' is a misnomer.

Coal-fired power stations are major emitters of carbon dioxide: but improved technology can reduce emissions by up to 20 per cent.

One option is to use steam at high pressure to drive turbines (advanced super-critical steam cycle, or ASCSC) in a coal-fired power station. This can increase efficiency to 45 per cent. Alternatively, heating or gasifying coal enables the carbon dioxide to be removed before combustion. Because this so-called integrated gasification combined cycle separates out the carbon dioxide, it is particularly suitable for adapting for carbon capture. Another option is to modify power stations to burn more biomass alongside the coal. This can allow 20-30 per cent of the fuel mix to come from biomass, bringing down emissions.

what potential does new coal have?

Globally, there are vast resources of coal, but without clean technologies burning coal to make electricity could come at a very high price to the climate. Approaches such as the integrated gasification combined cycle have not yet been tested on a commercial scale. Tougher pollution controls will reduce coal-fired electricity-generating capacity in the UK by about 30 per cent from current levels by 2016.

Friends of the Earth says any new gas- and coal-fired power stations should use the most efficient technologies, including the ability to use heat for local heating schemes and capture and store carbon emissions.

subsidised energy

The cost of developing renewable energy is often compared unfavourably with the costs of coal, oil and gas. What is often forgotten is the level of subsidy that the fossil fuel sector gets.

BP, Shell and Exxon – the world's largest oil companies – get public money from the World Bank and other institutions to fund new schemes, including pipelines and drilling projects.

The World Bank's mission is to reduce poverty. The UK government contributes funding in the form of 'development aid', paid for by the tax payer. Between 1992 and 2005, the Bank gave US $28 billion in loans for oil and gas projects – 17 times more than it invested in renewable energy. Despite growing calls for a shift to renewables the Bank continues to back fossil fuels.

Oil companies also get government-backed guarantees for risky developments overseas. This comes from export credit agencies, and is essential if companies want to attract private money in risky ventures. When BP wanted to build a pipeline from Azerbaijan to Turkey, it asked the UK's Export Credit Guarantee Department for help: if any of the countries involved in the project failed to deliver, the UK government would take on the debt. BP also got US$135 million from the World Bank to help build the pipeline.

There's more. Fossil fuels also get tax breaks and research funding from governments. Fossil fuel subsidies effectively increase greenhouse gas emissions and pollution.

nuclear power

Conventional nuclear technology uses fission – splitting atoms of heavy metals, such as uranium, to generate heat. This is used to produce steam to drive turbines generating electricity.

One drawback of nuclear energy is the waste product. Uranium fuel rods remain radioactive for thousands of years after they've stopped producing power. No safe way of disposing of nuclear waste has yet been found.

Given the serious consequences of accidents involving nuclear material, the safety record of the industry is a key consideration. Many think its record is poor – the most infamous accident being the explosion at Chernobyl in the former Soviet Union in 1986 that led to radioactive contamination across Europe. Scientists say it is still not possible to calculate the full extent of the damage but thousands of people are thought to have died as a result of contamination and 20 years later an exclusion zone almost as big as Greater London surrounded the plant. Safety concerns are also aroused by the fact that nuclear fuel can be used for

How a nuclear power station works

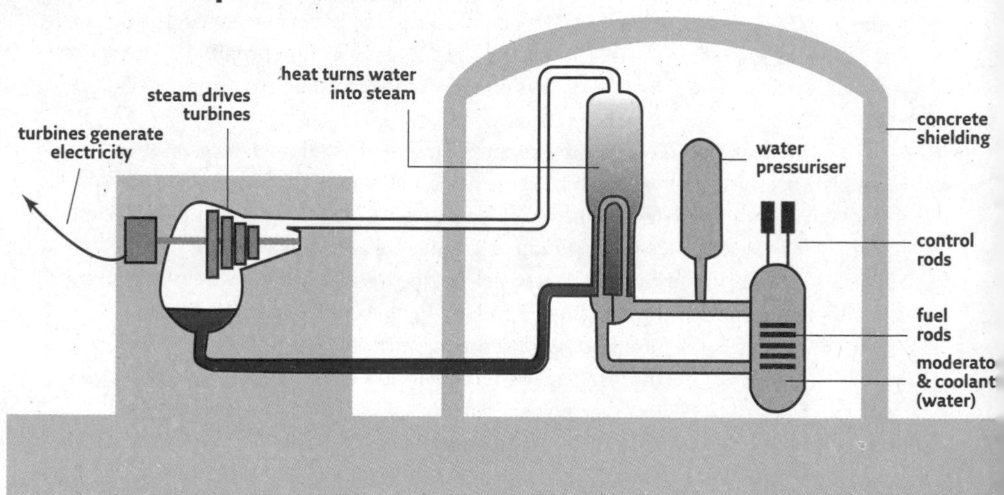

weapons. Britain's reactors have relied in the past on uranium from Australia. There have been reports of uranium going missing in transit, and it is impossible to guarantee that it won't end up in the hands of terrorists.

Experts say the safety features of modern nuclear power stations have greatly improved. Nuclear technology has advanced since the UK first invested in 1956. If a new programme goes ahead, it is likely that European pressurised reactors would replace the current advanced gas reactors and one remaining Magnox. More efficient designs still, such as pebble bed modular reactors, are on the drawing board.

Work is under way on a European pressurised reactor at Olkiluoto in Finland but two years into the build, it was already behind schedule with massive increases in costs.
No electricity is likely to be generated from the Finnish plant until at least 2011.

Nuclear power at a glance

Fuel	Uranium – limited supplies
Land requirements	Small
Lifespan	40 years
Pros	Lower emissions than fossil fuels. Potentially large quantities of baseload electricity from a small number of plants.
Cons	Many risks: radioactive waste with no existing disposal facilities; accident risk; radioactive emissions; terrorism threat; risk of nuclear proliferation. Direct competition for baseload with combined heat and power. Industry won't go ahead without huge public subsidy for decommissioning.

OVER-REACTION?

While there are big questions over the safety and security of nuclear power, its real contribution to the overall energy mix tends to get exaggerated: in fact it accounts for just 4 per cent of total energy consumption.

putting a price on nuclear

The economics of nuclear power are as controversial as the technology. Once billed as likely to become 'too cheap to meter', in fact costs have stayed high. New plants have tended to run into building delays and cost far more than their original budget. Costs at Torness in Scotland, completed in 1988, soared from £742 million to £2.5 billion. Sellafield's Thermal Oxide Reprocessing Plant was expected to cost £300 million but finished five years behind schedule with a final bill of £2.8 billion. In 2003 UK taxpayers bailed out the nuclear power company British Energy to the tune of £5 billion.

Uncertainties over the economics mean nuclear power is not an attractive option for private investors unless the government provides some form of guarantee.

what potential does nuclear power have?

Nuclear power today supplies 18 per cent of the UK's primary electricity supply – though only around 4 per cent of our overall final energy consumption. Britain's reactors are reaching the end of their lifespan. Seven power stations are scheduled to close by 2023, leaving just one in operation – at Sizewell B on the east coast.

The decision over whether to build new reactors has been controversial, but concerns over security of supply and the need to cut carbon dioxide emissions have been used to justify a re-consideration of nuclear power.

The UK government has not put a figure on how much electricity will come from nuclear power in a future energy mix. The potential depends on the willingness of the industry to build reactors, and the willingness of the public to allow them. Several companies are keen to build a new generation of plants, but want guarantees from government before they do so. Any new sites in the UK would face the challenge of public opinion. Under new legislation local communities are unlikely to have much influence over decisions about big energy projects, but politicians will still need to persuade the public that nuclear power is the right way to go.

According to IPCC experts nuclear power could potentially provide 18 per cent of the world's electricity, substantially reducing carbon dioxide emissions. But they point to problems with waste, safety issues and the threat of weapons proliferation as reasons why nuclear power might not realise its potential.

An alternative and supposedly safer source of nuclear energy is fusion, which merges atoms to create energy. Fusion is what makes the sun shine. Decades of research has so far failed to make fusion a viable option for electricity supply.

Britain's nuclear future – recommended locations for new nuclear power

In 2006 a government-commissioned report gave the strongest recommendation to these sites for new nuclear power stations:

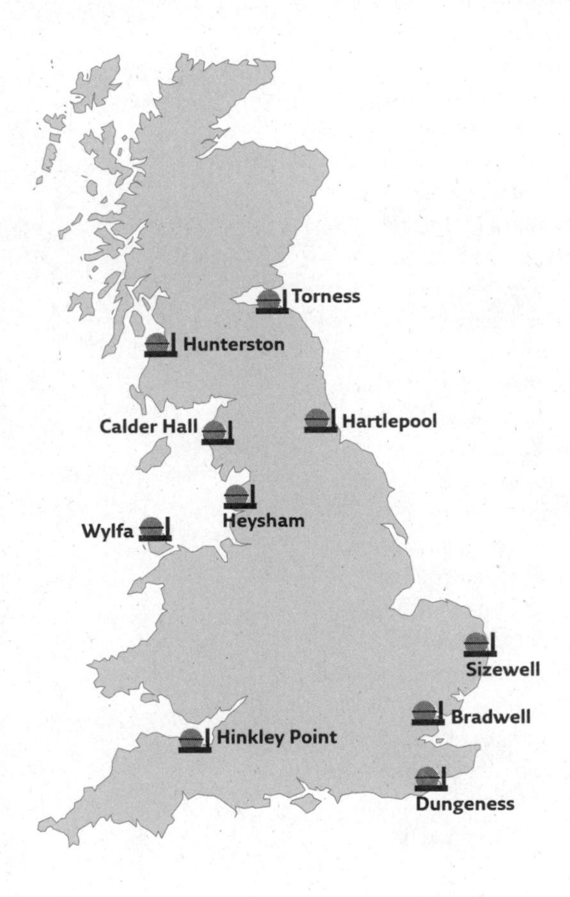

RENEWABLE SOURCES OF HEAT

There are lots of ways to reduce our carbon footprint from generating electricity, but more than a third of the carbon dioxide we produce results from generating heat for our homes, offices and factories – primarily using gas.

So what are the alternatives? Combined heat and power technology, where the heat given off in the electricity generating process is captured and used, can reduce the need for heating fuel – with power stations feeding excess heat into local heating systems. Wood-fired stoves, for example, can use local waste wood for heat and hot water in homes; and combined heat and power boilers can heat commercial and public buildings such as libraries and swimming pools. Small-scale wind can be used to heat water – a good way to store energy. Biofuels can replace conventional heating oils in homes that are not on the gas grid.

ground source heat

Ground source heat at a glance	
Fuel	Electricity (to run a pump)
Land requirements	Outside space needed
Pros	Useful for space heating
Cons	Not truly renewable, requiring energy to drive the pumps.

The ground absorbs and retains heat from the sun. During cooler months, temperatures underground are warmer than at the surface. Ground source heat pumps harness this energy using a coiled tube laid in shallow troughs or in deep boreholes. The tube is filled with liquid, which is pumped underground and

cornish toasty

When Penwith Housing Association decided to upgrade the heating system on the Chy An Gweal Estate in Ludgvan, Cornwall, in 2004, ground source heat pumps seemed to be the way to go.

The 14 bungalows on the estate, built in the 1960s, had no access to mains gas and had relied on coal-fired boilers, which meant hard work for the mainly elderly residents, as well as high fuel costs and high carbon emissions.

Denys Stephens, Senior Architectural Assistant at Penwith Housing Association, says ground source heat was attractive because it was low-carbon, cost-effective and feasible to install, especially as funding was available from the Clear Skies Programme and Penwith District Council. 'Heat pumps are not by any means zero carbon,' he says. 'But, for the amount of heat energy you get out of it, given the available options locally, it was about as low a carbon option for heating as we could get.'

Penwith Housing Association had already used ground source heat pumps at new-build developments in Cornwall, but Chy An Gweal was the first social housing project to have the technology installed after it was built. Each property, which had recently been refurbished to improve the level of insulation, was fitted with its own heat pump, with a plastic pipe fed down two 40 metre-deep bore holes dug into each garden.

To make the most of the heat supply, high water-content radiators were chosen for the heating system. The heat pump also supplies a hot water cylinder, providing hot water for each home.

'Infinitely better' is how one resident described the results. All supported the decision to install the system, which has meant a drop in fuel costs, with an average bill for heat and electricity for a two-bedroomed bungalow falling from £980 to £415. And because the heat pump generates more than three times as much energy as it uses, carbon dioxide emissions have fallen by around 70 per cent as well.

The success at Chy An Gweal has inspired other social housing providers to follow suit, with some 700 ground source heat pumps now installed around the UK.

back into a heat pump. This acts a bit like a refrigerator extracting and condensing the heat, which is used to warm water to around 40-50°C – good enough for central heating and hot water.

Although the pumps use electricity, they generally produce three to four times more energy than they use – so achieving a net carbon reduction. With a green source of electricity for the pump ground heat can be a completely carbon-neutral source of heat.

what potential does ground source heat have?

Ground source heat pumps are typically used for individual homes or commercial or public buildings. Space is needed outside to lay the pipes underground. A typical domestic system would cost £6,000-£10,000.

In commercial buildings heat-pump technology is more commonly used to provide cool air systems, with heat extracted from the air supply. But the technique is popular for heating in the United States and parts of Europe.

There's lots of potential for ground source heat pumps, particularly in new-build properties, where installation is cheaper. The Government estimates that around 1,550 large industrial sites in the UK could use them. On average each heat pump could generate 800 kW of thermal power.

In some parts of the world pockets of hot water and rock underground provide a natural source of hot water or warm air. These geothermal energy sources, formed by magma rising close to the Earth's surface, were used by the Romans for baths and today provide hot water and can be used to power steam turbines to generate electricity. Worldwide, there is some 7 GW of geothermal electrical capacity, with geothermal power stations in the United States, Iceland and Japan. Many countries in volcanic areas use geothermal hot water – in Oregon, United States, it is run in pipes beneath the roads to prevent freezing in winter.

There is only one geothermal plant in the UK – in Southampton, where it provides heat for the local hospital, office buildings and hotels. Underground hot water is found in some other areas, and is used for spa resorts, but there is limited potential for exploiting this commercially.

warmth from wood in Barnsley

Barnsley Metropolitan Borough Council has cut carbon dioxide emissions by 40 per cent after switching to wood chip as a fuel in some of the council's former coal-fired boilers. Wood pellets – produced partly from waste wood from the city's parks and trees – fuel a district heating scheme for council flats, as well as firing boilers in schools and council buildings. The council's new Westgate Plaza offices run on a wood pellet boiler, providing heat during the day for around 700 office staff. Heat produced overnight is stored for later use in the nearby library and town hall. Barnsley's new digital media centre will be heated by a wood pellet boiler and wood pellets will also heat new schools scheduled to be built in 2008.

Switching to wood has been 'win, win, win' according to designer and energy engineer at the council, Dick Bradford. 'It's a cheaper fuel, it's cleaner, it's carbon neutral, there are health benefits, and it's creating jobs and bringing woodland back into management,' he says.

In 2004 the council pledged to use biomass for any new boiler it installed provided the economic case could be made – and Dick Bradford says the economic argument for wood is difficult to ignore. Residents use heat meters to pay for their fuel. According to the council this has helped to reduce energy use as well as saving money.

Wood pellets, which can be used in coal-fired boilers, at first had to be imported from mainland Europe; but the council successfully lobbied for a local pellet mill – which will supply the rising demand from across South Yorkshire, where Sheffield and Doncaster councils are following Barnsley's lead. According to the South Yorkshire Forest Partnership, some 450,000 tonnes of local wood are available, providing the potential for even more energy. Dick Bradford estimates the council will hit its target of a 60 per cent reduction in carbon dioxide emissions by 2010.

five top green energy technologies

1. Offshore wind
2. Onshore wind
3. Solar (including concentrated solar power)
4. Sustainable biomass combined heat and power
5. Ground source heat

solar hot water

Solar hot water at a glance	
Fuel	Sun
Land requirements	Can fit on existing roof space
Lifespan	25 years
Pros	Easily retrofitted to existing housing stock. Can generate 50-80 per cent of heated water for an average house. Easily incorporated into new-build.
Cons	Requires a hot water cylinder, replumbing and a bigger tank in older stock.

As well as making electricity, the sun's heat can be used to warm water up to 80°C in homes and commercial buildings. Solar hot water systems – sometimes referred to as solar thermal systems – are usually fitted on a roof with a series of tubes filled with water collecting heat in sunlight. These tubes connect to a hot-water storage tank.

A range of technologies is available to collect the sun's heat, and typically up to 40 per cent of the heat is transferred to the hot water system. A system that turns the solar panels so that they get the most available sunlight can improve efficiency on large installations. In the UK a solar thermal system can provide hot water for a household's entire needs during the summer months, and can usefully pre-heat water in winter, saving a lot of fuel. A typical domestic system can cost between £2,500 and £8,000 and save half a tonne of carbon dioxide a year from the average household.

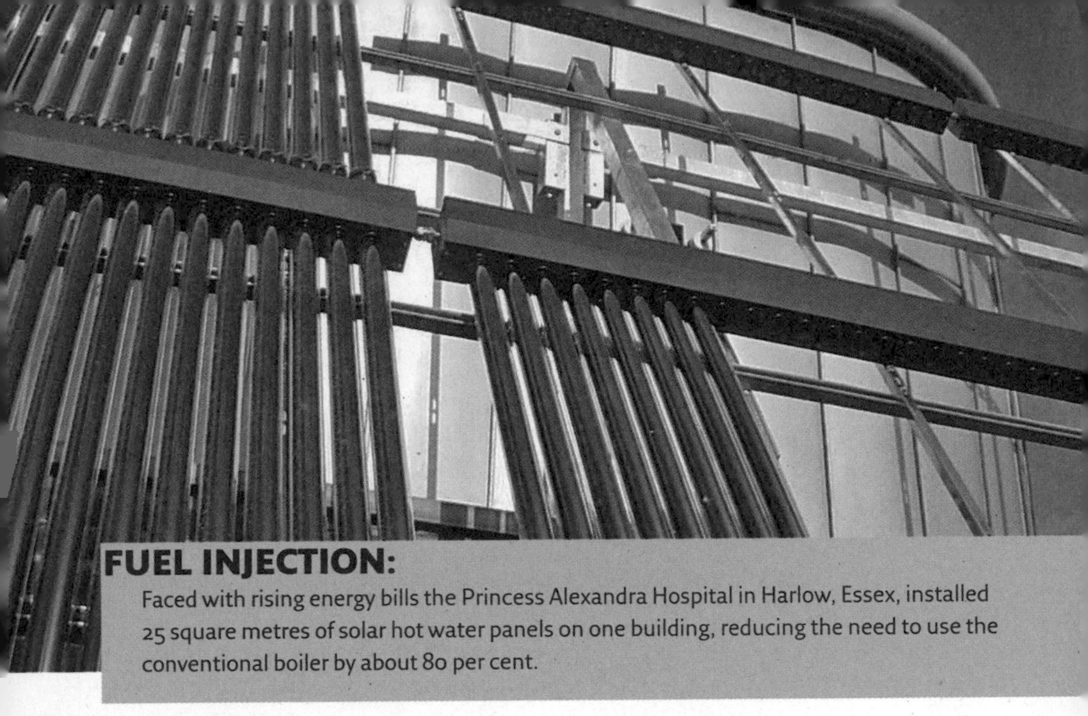

FUEL INJECTION:

Faced with rising energy bills the Princess Alexandra Hospital in Harlow, Essex, installed 25 square metres of solar hot water panels on one building, reducing the need to use the conventional boiler by about 80 per cent.

what potential does solar hot water have?

The technique is increasingly being used in homes – around 10,000 solar hot water systems are now installed every year. It can also be used for commercial buildings and factories, providing hot water for swimming pools and industry. In fact, any commercial or public building that uses a lot of hot water can benefit from solar hot water.

Barcelona's solar by-laws

Under a special by-law, introduced in 2000, all new buildings and any undergoing substantial renovation in Barcelona must now be fitted with enough solar panels to provide 60 per cent of the hot water. There are now more than 40,000 square metres of solar panels in the city, saving more than 5,000 tonnes of carbon per year. Other cities across Spain are introducing similar by-laws.

local versus central?

Britain's energy system is centralised: most people are on the National Grid and mains gas. But many low-carbon energy alternatives don't need to be hooked into the grid – they can provide power to just one local building or a neighbourhood. Decentralised systems can get more out of the fuel we use because less energy is lost in transmission, and that should translate into lower greenhouse gas emissions. Although maintenance costs may be higher for a decentralised grid, power losses are lower, especially at times of peak demand. And greater efficiency saves money.

A centralised grid like the UK's has the added advantage of being able to provide lots of back-up power to cope with sudden spikes in demand, or an abrupt loss of power from an individual plant. This can be particularly important for users who rely on variable sources such as wind or solar PV to meet some of their needs: when the wind stops blowing or when night falls they can turn to the grid for back-up. A large grid also has more capacity to incorporate major new energy supplies, such as the levels of power generated by concentrated solar power. Some experts have advocated introducing high-voltage direct current (HVDC) cables to create a Europe-wide grid.

An intelligent grid could build on the advantages of both local and central supply. This would link power from a vast range of sources, including local energy systems, into an energy web using computers to balance supply with demand. When demand is high, more systems could be automatically brought into the grid – and when demand is low, supply reduced.

a supergrid

Proposals have been put forward to link Europe and North Africa with a supergrid, using high-voltage cables running under the Mediterranean, Irish Sea, North Sea, Baltic Sea and English Channel as well as over land. Electricity generated across Europe and beyond – from offshore wind in the Irish Sea to concentrated solar power in southern Europe and North Africa – would feed into the grid, providing a common resource across the continent.

Such a grid requires special high-voltage cables that would make it possible to carry a high load and link to the national grid systems of individual countries. High-voltage supergrid cables already form part of the UK's National Grid.

A pilot project linking a wind farm in the North Sea to the UK, Germany and the Netherlands has been proposed. For it to go ahead the scheme will need political support from across the European Union and agreement on a common energy policy as well as financial investment. These requirements currently represent a major obstacle.

RENEWABLE TRANSPORT FUELS

While heat and electricity account for roughly two-thirds of our energy use, transport takes up most of the rest: around a third of our carbon dioxide emissions come from aeroplanes, trains and automobiles. In 2005, road vehicles account for more than two-thirds of UK transport emissions (if we include international aviation).

	million tonnes CO_2
Road transport	119.9
Other transport	2.4
Aviation (incl. international)	37.5
Shipping (incl. international)	10.0

As we saw in Chapter 4, improved transport planning, better fuel efficiency and other measures have the potential to reduce the level of demand for fuel. But we have to get around. So what alternatives are there to petrol and diesel to power our lorries and cars? And what about aviation, the fastest-growing source of climate-changing emissions?

road transport

Most options for curbing emissions from road transport focus on the amount of fuel a vehicle uses. Few carbon-neutral options are available, although manufacturers have developed prototypes. One study estimated that with the right policies promoting a switch to lower-carbon cars we could halve emissions from road transport from 1990 levels. A switch to biofuels, liquid petroleum gas, compressed natural gas (CNG), hydrogen and electricity has been estimated to be

GET A MOVE ON:

Time to tackle the climate-changing emissions – about a third of the total – coming from cars, aeroplanes and boats.

capable of saving up to about 9 million tonnes of carbon. The International Energy Agency has said that 20–40 per cent of all transport fuels could come from alternative sources. In the UK 5 per cent of vehicle fuels will come from renewable sources by 2010, following government legislation.

The UK Government wants to bring road transport under the remit of the Emissions Trading Scheme (see p.224) as a way of cutting pollution. Suppliers could be required to hold carbon allowances for all the fuel they sell; but as with emissions trading in other areas, the success of the scheme will depend on the quotas provided and the resulting cost of carbon. If fuel suppliers are given large emissions quotas, the price of carbon will remain low and there will be little incentive to cut emissions.

biofuels

Cars, buses, trucks and motorcycles can use liquid fuels made from organic material such as plants, animal waste and sewage. Vegetable oils – notably from soy, palm and oilseed rape – can be turned into biodiesel; starchy, sugary vegetables like corn can be used as ethanol (petrol substitute), and animal and vegetable fats can be recovered from the food industry and converted to biofuel.

Biofuels have been promoted as a solution to climate change because the plants absorb carbon dioxide when they grow and release it when they're burnt. The European Union has set targets to increase their use as transport fuel. But meeting these targets throws up some serious social and environmental and challenges.

Growing, processing and delivering transport biofuels uses lots of energy and artificial fertiliser. One study by a Nobel prize-winning scientist suggested emissions from the most commonly produced agrofuel in Europe – biodiesel from rape seed oil – could be up to 70 per cent higher than those from conventional fossil fuels.

PALMED OFF:

Biofuels like those made from palm oil are a phoney solution to climate change – they're leading to the trashing of tropical forests and communities living in them.

Other problems are associated with the impact of farming. Meeting the EU targets would rely on large monoculture plantations (ie where only one kind of crop is grown), and these can be massively damaging locally, as well as being a significant source of greenhouse gas emissions. Biofuels produced in this way are often referred to as 'agrofuels'.

Because there isn't enough farm land available in Europe, meeting the EU targets would mean importing agrofuels, mostly from developing countries. Malaysia and Indonesia, already two of the largest oil palm producers are gearing up to feed the European agrofuels market. Palm plantations are usually set up on land cleared from tropical rainforest and peat bogs, causing massive carbon dioxide emissions. In fact this is the reason Indonesia is now the third biggest source of greenhouse gases after China and the United States. Palm oil expansion is threatening to destroy the last remaining habitat of the orang-utan, man's closest genetic relative, and forcing people off their land. Africa's remaining forests could also be at risk.

One effect of turning land over to agrofuel production is to drive up food prices, often affecting the world's poorest people worst. Peasants in Mexico have protested against price rises in maize – a staple food – partly as a result of the United States diverting maize production to agrofuels. Sugar cane, largely from Brazil, is one of the agrofuels with the most climate benefits. But its cultivation has led to massive land clearance, and labour conditions on plantations are often appalling. Extension of soya plantations, especially in Latin America, is fraught with social and environmental costs – tropical deforestation leading to climate change, loss of biodiversity and collapse of people's livelihoods.

The biofuel industry is promoting a plant called Jatropha as a miracle crop because it can grow on poor-quality land in hot climates. But it grows even better on good farm land: already people are being encouraged to convert to Jatropha and it will therefore compete with food production.

the potential of biofuels in transport

The true potential for biofuels is unclear but they probably represent only a small part of the solution to our energy needs. For them to help make significant cuts in greenhouse emissions, and to ensure the industry does not do more harm to the environment and communities than the fuels it is replacing, would require strong, mandatory rules about the way the fuels are sourced. It is difficult to see how they can ever be produced on the scales advocated by the EU without damaging the environment and communities.

Friends of the Earth believes agrofuels should not be considered part of the solution to climate change. The organisation recommends a ban on the import of biomass and biofuels to the EU, and for a moratorium on all bio-energy targets and subsidies.

If done in a sustainable way first-generation biofuels can help reduce emissions by a tiny percentage, well below the targets set by the EU. If done unsustainably they are likely to add to the problem of climate change.

hybrid vehicles

Hybrid vehicles have a conventional internal combustion engine alongside an electric motor that is powered by a rechargeable battery. At low speeds the car runs on the battery, switching to the petrol engine as speeds increase, with energy from the engine charging the battery at the same time.

Petrol-electric hybrids were the first to come on to the market, although diesel-electric versions are also being developed. There is a growing number of hybrid vehicles on the market, with the Toyota Prius the best-seller in 2007. Carbon dioxide emissions from the Prius are 104 g carbon dioxide/km compared to an average of 167.2 g/km for cars sold in the UK in 2006. Hybrid vehicles may be more expensive than standard models; but as the market grows, prices could come down.

A dual fuel system means that hybrid cars use less fuel than petrol-only vehicles and on average reduce emissions by around a quarter.

electric cars

Electric cars use rechargeable batteries to store the power they need. The battery is charged partly by the car's braking system, and partly from mains electricity.

The amount of emissions from electric cars depends on the source of electricity used to charge the battery: renewable sources can provide carbon-free driving, but even using electricity from fossil fuels offers a 40 per cent cut in carbon dioxide emissions.

Batteries can provide only a limited store of energy, making electric cars suitable for short distances at low speed. GoinGreen's G-Wiz has a maximum range of 48 miles and a top speed of 45 mph making it ideal for commuting. Charging from the mains for a small car like the G-Wiz takes eight hours. The G-Wiz has energy-saving low-resistance tubeless tyres, energy-absorbing bumpers and is officially classified as a quadricycle, not a car.

scooters for the climate

The Vectrix Maxi-Scooter is a light aluminium-framed scooter that runs on a battery. With a top speed of 62 mph and zero emissions if charged from a renewable source, it has a range of 75 miles. Vectrix, which set out to provide clean efficient transport using two-wheeled vehicles, has also developed a hybrid scooter, which can go 155 miles before needing a re-charge.

hydrogen

Hydrogen is not a fuel source but a way of storing energy. It has been used for space travel, but the technology has not yet been mastered for cars.

BMW is just one of the manufacturers exploring the potential of hydrogen. In 2004 it set speed records with a specially-designed hydrogen-powered racing car. Its 745h model, released in 2006, has a conventional engine that runs on either petrol or liquid hydrogen. With most hydrogen made from fossil fuels, the car's carbon emissions are probably higher than the average vehicle on the road.

Hydrogen can be produced by burning coal and gas, releasing carbon dioxide, or it can be made from the electrolysis of water. Some algae and bacteria produce hydrogen in sunlight and studies to investigate how these sources could be tapped are still being developed.

Hydrogen is a tricky way of storing energy. Extremely volatile, it must be stored as a liquid, requiring a temperature drop to -253°C or storage under high pressure.

Its drawbacks may explain why hydrogen is not generally available at your local filling station. In Germany hydrogen pumps are beginning to be seen more and a hydrogen fuel system is planned for buses in Beijing for the 2008 Olympics and for Shanghai's taxis by 2010.

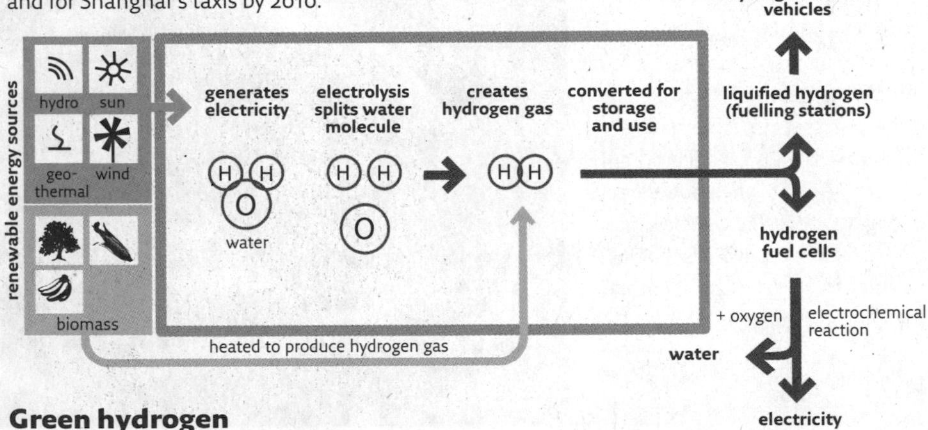

Green hydrogen

Most car manufacturers are focusing on developing hydrogen fuel cells rather than using hydrogen to power conventional car engines. Fuel cells produce electricity by combining two fuels – commonly hydrogen and oxygen. The resulting power drives the vehicle's engine, producing water as a waste product. Fuel cells are efficient and, provided the hydrogen comes from a clean source, could reduce carbon dioxide emissions to zero. Fuel cells need a ready supply of hydrogen if they are going to work for vehicles. They are not yet in the showrooms, although there has been more progress with using them in larger vehicles such as buses.

Realistically, hydrogen is not going to be available commercially for the mass market for perhaps 20 years. We don't have that long to tackle climate change. Hydrogen is a longer term option.

vision of a hydrogen economy

Hydrogen enthusiast and economic expert Jeremy Rifkin says the world should be preparing for a new energy era: 25 years from now, he says, we will rely on renewable energy, stored as hydrogen. Every home, every office, every shopping mall will produce its own energy and store it in fuel cells.

The world will be linked by an energy internet – an open access grid where we can offload the surplus we generate or download extra power when required. 'We are talking about changing the whole infrastructure of the planet in 25 years,' he says. 'We have to get past carbon.'

Our homes and our cars will be our personal power stations – generating electricity from wind turbines, solar power and combined heat and power and storing it in hydrogen fuel cells. When we don't need the energy, it can be fed back into the grid. Hydrogen cartridges with five times the power of lithium batteries will mean we can charge laptops, music players and mobile phones.

The biggest beneficiaries, says Rifkin, will be countries in the developing world, where one third of the population currently has no access to electricity. Fossil fuels provided the elite and powerful with energy – but renewable power will be available for everyone.

In Huesca, Spain, a research project is using solar PV and wind power to produce and store hydrogen. Supported by Aragon's local government, the project is exploring the best methods of integrating the technologies, using hydrogen for generating electricity in combined heat and power, as well as using it in fuel cells for transport.

The European Parliament has called for a hydrogen economy in all European states by 2025, providing power for transport as well as electricity.

hydrogen on the buses

In 2005 Transport for London took part in a European trial, running three London buses on hydroge cells. Buses can be more easily adapted to run on hydrogen because there is room to store the fuel and refuelling can be slotted in to the bus's regular schedule.

Specially-built buses were fitted with nine cylinders storing more than 40 kg of compressed hydrogen on the roof. The buses, which had a top speed of 50 mph and could travel 250 miles on fu tanks, proved a success.

Eight other European cities also took part in the trial, with hydrogen-fuelled buses operating in Amsterdam, Barcelona, Hamburg, Madrid, Porto, Luxembourg, Stuttgart and Stockholm.

aviation

Aviation is the fastest-growing source of carbon dioxide. Official figures attribute about 6 per cent of UK carbon dioxide emissions to aviation in 2005, although because of the altitude of the emissions their impact on climate change is twice as great. Airlines currently rely on kerosene – an oil-based fuel – to get them off the ground, but what alternatives might be available in future? The options are similar to those for road transport – but flying 13 km high can create other difficulties as well.

flying on biofuels

Biofuels are currently not a suitable aviation fuel because they cannot cope with the cold in the upper atmosphere. Scientists are investigating how biofuels can be adapted and mixed with kerosene. Sir Richard Branson has said Virgin will fly a 747 on biofuel in 2008 and that in the long term the entire airline fleet would run partly on biofuels. If biofuels were to prove a viable aviation fuel, however, the increased demand would put agricultural land under even greater pressure (see Biofuels, above). Some argue that biofuels might simply be used as an excuse not to address the underlying problem with aviation, which is its rampant growth.

flying on hydrogen

By contrast with biofuels, liquid hydrogen has already been used successfully in aviation, replacing kerosene in one of a plane's internal combustion engines. But switching commercial airlines to hydrogen is still some way away and most experts say there is in fact little potential for using hydrogen for commercial airlines. As with road transport, the two main problems are producing and storing the hydrogen: it contains less energy per unit of volume than kerosene, so a plane would need to carry far greater quantities, requiring an enormous fuel tank.

Reducing emissions by using alternative airline fuels seems, in fact, to offer very little potential for cutting our greenhouse gas emissions in the short term.

the future energy mix

Given the available alternatives, what could the UK's future energy mix look like in a low-carbon future?

A medium-term vision is offered by the government's Energy White Paper. This suggests that by 2020 efficiency measures could have reduced UK electricity consumption from 368 to 352 TWh – a reduction of just 4 per cent; while renewables are expected to increase to 67 TWh – 19 per cent of our total electricity. The overall effect would see carbon emissions reduced by up to a third.

Looking further into the future, one study suggests that by 2050, micro-generation – small-scale renewable energy at a domestic level – could provide 30-40 per cent of the UK's total electricity needs.

The government scenarios do not make cuts on the scale that climate science says is necessary. An alternative route map, produced for Friends of the Earth, has wind, solar and marine energy providing half of the UK's electricity by 2050, with carbon capture and storage taking care of the carbon dioxide emissions from fossil fuels, accounting for another 38 per cent. Energy-saving measures mean a reduction in total demand, with some of the electricity used to produce hydrogen for transport fuel. Carbon dioxide emissions would fall by 90 per cent on 2004 levels.

A study of the effects of adding a combination of wind, wave and tidal power to the electricity grid found that the mix of energies would provide a far more reliable source of energy than any single source. The mix would reduce peaks and troughs in supply and raise capacity at times of high demand.

There are many energy options, from local power to imported solar, smart grid electrics to travelling less. No one solution will cut our emissions enough – it is the package of solutions that must be embraced if we are to cut them at all. The choice is ours.

there is no single silver bullet, but many

Research for Friends of the Earth by the Tyndall Centre for Climate Change at the University of Manchester suggests that our energy supply needs to come from a mixture of different sources: renewables – especially wind, wave and tidal power – have lots of potential, and with the right support they could massively curb our greenhouse gas emissions. Other approaches such as solar hot water and getting heat from the ground offer savings to households; and biomass, such as wood, is useful where there is a ready supply of sustainable fuel. All of which could be developed while we make the transition away from fossil fuels. And to make the transition as cleanly as possible we need to improve the way we use them – for example, through so-called clean coal technology and by capturing carbon rather than simply pumping it into the atmosphere.

Low-carbon sources may increase the cost, but will offer benefits, including more jobs and better air quality, as well as being kinder to the climate.

how we can make change happen

How quickly can we achieve a low-carbon life? Here are some realistic approaches to help us make the switch.

chapter 6

bring on the eco age

Hailed by some as the world's first eco-city, Dongtan in China is not just a show-piece. This planned fossil fuel-free development on Chongming Island near Shanghai is meant to be a proper city for 80,000 people to live, work, play, and go to school.

Dongtan is a vision born of necessity, says Peter Head, planning director for the design and business consultancy Arup, whose team is behind the plans. China's economic boom has had a crushing impact on the environment. Air pollution in its cities is a big health threat; water and land are getting dirtier; and of course China recently claimed the dubious tag as the world's largest emitter of carbon dioxide. 'The industrial model that the West invented does not work,' says Peter Head. 'It pollutes, it weighs too heavily on the planet. This is understood at the highest level of government in China. They know it is really serious – it is a case of needs must.'

Local government targets for economic production have driven China's growth nationally, but now new targets are being considered for reducing pollution and waste. 'China, at the highest level, talks about this as a transition to the ecological age of civilisation,' says Peter Head. 'They talk about this as a revolution.'

Perhaps China, which has so recently jumped on the consumer bandwagon, can provide a model for the rest of us. Dongtan sets out to show that, with a clean sheet, we can create a low-carbon way of life that still meets our needs and comforts. The priorities shaping the city are the same the world over, and the technology and design principles in use there are

“ It is possible to 'decarbonise' both developed and developing economies on the scale required for climate stabilisation, while maintaining economic growth in both.

The Stern Review

> **“ The historical experience of human progress shows that we should never seek development at the cost of wasting resources and damaging the environment.**
>
> *Hu Jintao, President of the People's Republic of China*

beginning to spring up elsewhere. So what are the big steps we need to take together to bring the many solutions and brilliant ideas to life – while making sure the lights stay on and we get around?

Schemes to stop greenhouse gases reaching catastrophic levels range from dumping iron filings in the ocean to personal carbon allowances, from concerted government intervention in the economy to 'the market will decide', from wholesale changes in individuals' behaviour to the collective efforts of small towns.

This chapter takes a look at the most realistic approaches to setting us on the path to a low-carbon life. What will make the difference quickly enough and in a way that is fair to people and the planet?

FLOAT YOUR BOAT?

Time for political leaders to stop flirting with solutions and get on with implementing them.

the world's first eco-city?

Chongming Island sits in the mouth of the Yangtze River, less than 40 miles from the booming centre of Shanghai. Its silty soil provides fertile farmland, and at the tip of the island lies a unique area of wetland, home to the black-faced spoonbill, and a conservation site of international importance. The city is separated from the wetland by a wide park, and has been designed to ensure that it does not affect the visiting birds: development will only be allowed if it does not pollute the wetland.

The aim is to provide a high quality of life that has a sustainable footprint – and this has shaped the design. The new city will have very low emissions: people will live in low-energy buildings drawing heat and power from renewable sources; they will compost or recycle their waste and most of the food they eat will be produced locally.

When complete, Dongtan will consist of three towns around a central city area and connected by canals and water ways. No fossil fuels will be used in the city. A central energy park will use rice husks as fuel in a combined heat and power plant. The husks will come from rice mills on the Yangtze River and will be transported by barge to Dongtan. Wind turbines and solar PV will provide extra electricity. Buildings will all be designed and insulated to use an average of a third of the energy of comparable buildings in nearby Shanghai.

Food production could be a key to making the city sustainable. The island's farms will be in the city. Human waste will be collected and treated to make nutrients for plant fertiliser and solid waste will be turned into fuel as biogas. Dongtan will use two separate water systems: one providing drinking water and the other grey water for toilets, washing machines and gardens. Waste water will be recycled. Buildings will have 'green roofs', with plants providing insulation, helping to absorb rainwater and encouraging wildlife. Planners hope to encourage more wildlife in general by including plenty of parks and gardens.

In this compact, high-density city people should have little need to drive. Everyone will live within seven minutes' walk of public transport – and hospitals, schools, parks and shops will all be nearby. Water taxis and hydrogen-powered buses will provide transport, and there will be a network of cycle paths and walkways. Visitors will be encouraged to use public transport. Any vehicle in Dongtan must be free of greenhouse gas emissions. Cars will run on electricity or hydrogen to cut pollution and traffic noise.

The city has also been designed to withstand rising sea levels and storms, with a flood defence wall and underground reservoirs to hold flood water. The network of canals and ponds will absorb heavy rainfall, while the wetland provides a natural barrier against storm surges and tsunamis.

Work has started on the first section of the city. Some 3,000 homes were expected to be complete in time for the Shanghai Expo in 2010.

bringing down emissions – triangle by triangle

At the start of the 21st century two US researchers put together a list of 15 measures, each of which would cut global emissions by 1 billion tonnes of carbon. The idea was that just seven of these measures would have to be implemented globally to stabilise the amount of carbon dioxide in the atmosphere (today we're pumping out about 7 billion tonnes of carbon a year).

Stephen Pacala and Robert Socolow envisaged each cut of 1 billion tonnes as one wedge on a graph, adding up to a cut of 25 billion tonnes of carbon over 50 years, and stabilising emissions at around current levels.

The triangular emission cut options, known as Socolow's wedges, are:

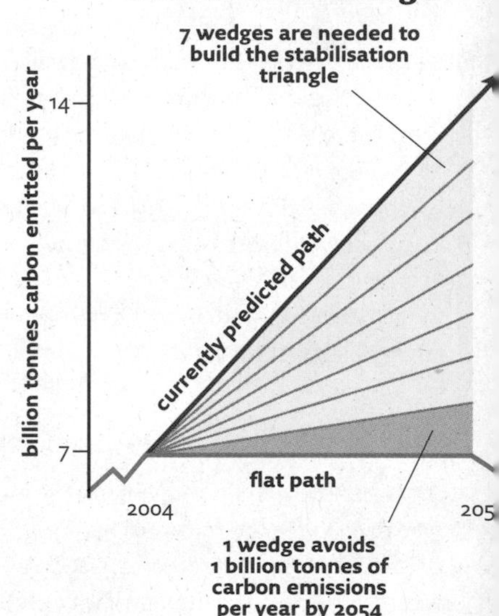

1. Doubling the fuel economy of 2 billion cars from 30 miles per gallon to 60 miles per gallon.
2. Halving the average travel for 2 billion cars from 10,000 to 5,000 miles per year.
3. Cutting carbon emissions from buildings by 25 per cent.
4. Increasing the efficiency of coal-fired power stations to 60 per cent (compared to 32 per cent today).
5. Replacing 1,400 GW of coal plants with gas running at 50 per cent efficiency, increasing gas power generation by a factor of four.
6. Capturing and storing the carbon at 800 GW of coal or 1,600 GW of natural gas plants within 50 years (and burn the hydrogen produced to make electricity).
7. Storing the carbon dioxide by-product from hydrogen plants by introducing CCS at plants producing 250 million tonnes of hydrogen per year from coal or 500 million tonnes of hydrogen per year from gas.

8. Capturing carbon dioxide produced by coal-to-synthetic fuel plants producing 30 million barrels a day of synthetic fuel.
9. Doubling the capacity of nuclear power to replace 700 GW of coal-fired power.
10. Adding 2 million 1 MW peak windmills – on- or offshore.
11. Increasing solar photovoltaics by 2,000 GW of power, replacing coal.
12. Using wind power to produce hydrogen fuel cells, by adding 4 million 1MW windmills.
13. Increasing the use of biomass for transport fuel, at 100 times the current rates in Brazil or the United States, using one sixth of the land available for crops.
14. Reducing tropical deforestation to zero and doubling the rate of new tree plantations – with either 250 million hectares in the tropics or 400 million hectares in temperate zones. A further half wedge could be achieved by planting 300 million hectares of new plantation on non-forested land.
15. Using 'conservation tillage' farming methods on all cropland – planting seeds by drilling holes rather than ploughing the soil, so reducing carbon dioxide loss.

Not every wedge is a real solution. Using a sixth of the available land for biomass production, for example, would have a disastrous effect on food production and prices and the poorest people in society would be worst hit. Relying on nuclear power would require unprecedented international security agreements and new measures to deal with highly toxic waste. The production of synfuels (synthetic fuel) from coal would require significant investment in the continued use of fossil fuel.

But Socolow's wedges do at least give a sense of the scale of change we need – and that there are lots of options to choose from.

" Humanity already possesses the fundamental scientific, technical, and industrial know-how to solve the carbon and climate problem for the next half century.

Pacala and Socolow, 2004, Science: 305: 968

contraction and convergence

There is widespread agreement that global emissions must be reduced in a way that is fair to rich and poor countries. But how do we make sure this happens?

Contraction and Convergence is an idea that attempts to reduce emissions fairly. 'Contraction' means bringing them down by agreeing a global budget for the total amount of carbon dioxide it's safe to have in the atmosphere. 'Convergence' means sharing that budget equally among all the people in the world. The aim would be for all countries to converge on the same allocation of carbon dioxide per person by an agreed date. Countries wanting more than their fair share would be able to buy unused allocations from others through an agreed trading scheme. Contraction and Convergence recognises the need for fair access to our atmosphere – a common global resource.

First put forward in 1996, Contraction and Convergence was developed by the Global Commons Institute (GCI). GCI argues that working towards a fair distribution based on population is the only way to secure global agreement on emissions reductions. The approach has won support from a number of countries.

Friends of the Earth agrees that the approach to tackling climate change should be fair and just. But it argues that the scheme should be modified to recognise the different needs of different countries. People living on a remote Pacific Island, say, may need a larger carbon allowance – so that they can fly to the nearest mainland, for example – than someone living in mainland Europe.

sustainable development – what is it?

'Development that meets the needs of the present without compromising the ability of future generations to meet their own needs.'

This was how the concept of sustainable development was described in 1987 in *Our Common Future*, also known as The Brundtland Report after its author, former Norwegian Prime Minister Gro Harlem Brundtland, in her role as Chair of the World Commission on Environment and Development. The agenda-setting report put the case that economic development must go together with social justice and protection of the environment.

The term 'sustainable development' has since been adopted by a wide range of people, politicians, organisations and businesses, many of whom have used it to fit around their own agendas. In 1997 the oil company Shell, for example, revised its business principles to include a commitment to contribute to sustainable development, while continuing to invest in pumping oil and gas – which many people believe is incompatible with that goal.

Although there is widespread disagreement about how the term translates into practice, most definitions of sustainable development revolve around the idea that people need economic progress that must go hand in hand with a healthy planet and social justice.

Bogotá, a city for people

'We had to build a city for people. Instead of building highways, we restricted car use. Public space is for living, doing business, kissing and playing. Its value can't be measured with economics or mathematics; it must be felt with the soul.'
Enrique Peñalosa, Mayor of Bogotá, 1998-2001

When Enrique Peñalosa was elected mayor of Bogotá, Colombia, in 1998 he asked why, given that most (85 per cent) of the people did not drive to work, cars took up so much space on the street. He started introducing policies to restrict cars in the city, banning parking on pavements, building cycle lanes, and excluding cars from areas of the city on different days of the week. The TransMilenio system, built in just two years, copies the Brazilian city of Curitiba with rapid transit buses running down the middle of the street carrying 780,000 people a day – and making a profit. By 2015 more than four in every five residents should live within 500 metres of a bus station. Residents have voted to outlaw cars from the centre by 2015.

greenhouse development rights

One approach to dealing with the need for global emissions reductions is to look at who can afford to pay for cuts to be made.

The idea of Greenhouse Development Rights – put forward by the US organisation EcoEquity and supported by UK development agency Christian Aid – is an attempt to tackle the need for emissions cuts alongside the need to tackle global poverty.

Because of the level of urgent emissions cuts needed, allowing equal per capita emissions for all countries would effectively condemn the world's poorest to a future of poverty. Instead, EcoEquity proposes allowing the poorest people the right to develop, with those who can afford to pay the costs of cutting emissions (including rich people living in poor countries), footing the bill.

Such a system relies on calculations that determine to what extent each country is responsible for global emissions (taking 1990 as a base level) and also looking at their financial capacity. From this, the proposed framework calculates who should be responsible for funding the measures needed to support countries suffering the effects of climate change as well as funding emissions cuts to ensure global temperature rise remains below 2°C.

Proponents of the scheme say it could be the only way in which developing countries can accept a global deal to cut emissions. For more, see www.ecoequity.org

greening India's transport

There was a time when a walk around the Indian city of Ahmedabad would have left your eyes and nose stinging from the pollution in the air. No longer. Today the city's public transport, taxis and rickshaws all run on compressed natural gas (CNG) – with unhealthy particulates some 60-97 per cent lower than from old diesel engines.

In fact Ahmedabad is one of 11 big Indian cities to have followed the lead of the capital Delhi – the first major city in the world to convert its entire bus fleet (some 12,000 vehicles) to CNG. Delhi's 14 million inhabitants have every reason to welcome steps to curb pollution: in 1998 it was estimated that one person died every hour because of poor air quality.

Whether or not CNG is the best fuel for everywhere, the case of Ahmedabad shows how much a city can do to curb the impact of its public transport.

The success of the Delhi scheme spurred India's Government to propose an Auto Fuel Policy requiring all new vehicles to meet European III emissions standards. The most polluting cities had to apply this regulation by 2005 with the rest of India having to fall in line by 2010.

cunning plans in the UK

An array of public authorities hold the key to driving forward climate-friendly practice around Britain – nationally, in the regions and for cities and neighbourhoods. One of their most important tools is the planning system – the way we decide how land is going to be used, what gets built and what doesn't.

In England responsibility lies with Regional Development Agencies and local planning authorities, which could be taking steps such as:

- obliging all new developments to meet the highest efficiency standards in energy and resource use
- throwing their weight behind renewable energy projects
- curbing developments with major carbon dioxide emissions
- making sure development reduces people's need to travel by car.

Wales, Scotland and Northern Ireland have devolved responsibility for planning. The Scottish Government has produced its own Climate Change Programme, and the Welsh Assembly has published an Environment Strategy and has its own planning policy.

Local authorities can develop sustainable energy strategies for their area and these could include district heating schemes and local electricity grids powered from renewable supplies. The London Borough of Merton has used its planning powers to require developers to include targets for renewable energy.

Given the urgency of the need to address our carbon footprint it's reasonable that at the bare minimum all councils should have policies that mirror those of the more progressive ones. They could start setting ambitious carbon reduction targets for their local area. In fact the government has said local planners must take action to ensure that the country delivers on its national and international climate change obligations.

There is huge scope for our public authorities, especially through planning, to lead the way in finding solutions to climate change.

case study

capital ideas

Almost half of the world's 6 billion people live in cities – and the proportion is growing. Cities are responsible for about four in every five tonnes of carbon dioxide emitted globally. They are often the driving force of an economy and so have an important role in leading change.

the greening of London ...

London's city structure and mayor are unique in the UK – and they have special powers to address climate change. London is benefiting from the lessons of Dongtan (see p.202). Peter Head, a sustainable development commissioner for London, says the two cities have more in common than you might imagine. 'Dongtan is just one island in China but it does have an ability to inform the way we deal with existing cities. London has a similar density – and you need that density to make public transport viable. It also has a similar number of green spaces,' he says.

Ways of putting these lessons to practical use appear in Mayor Ken Livingstone's London Climate Change Action Plan, which aims to cut the capital's carbon dioxide emissions by 60 per cent from 1990 levels by 2025; it has targets for cuts from existing homes, business, transport and new buildings.

The biggest single source of London's emissions is gas and electricity use, so this is a key area to address. The proportion coming from renewable supplies is set to double by 2025 if the Action Plan works. Local heat networks fuelled by gas, biomass or hydrogen fuel cells are to be introduced, providing power, heat and cooling systems. Some new developments will go ahead on the presumption that they will achieve reductions in carbon dioxide emissions of 20 per cent by generating their own renewable power.

Energy use in the home is responsible for just under half of London's emissions. As part of the Action Plan, free insulation will be available for social housing; and private landlords who improve the energy efficiency of their properties will be eligible for a green badge scheme showing tenants that energy efficiency has improved.

Public buildings, offices and shops are big energy users. If two thirds of them turned off their lights and photocopiers at night, they could save nearly 1 million tonnes of carbon a year.

Nearly a quarter of the city's carbon dioxide emissions come from transport. The Congestion Charge, which has to be paid by vehicles entering the city centre, with exemptions for low-emissions hybrid and electric vehicles, discourages drivers from the centre at busy times. Buses in the capital are being replaced with hybrid diesel-electric vehicles – and 10 hydrogen-fuelled buses will join the fleet by 2010.

To help it all happen, the Mayor of London Ken Livingstone said the government needs to use financial incentives to encourage companies and individuals to act. He called for a comprehensive system of carbon pricing and changes to regulations to improve the energy efficiency of new buildings.

The mayor is not acting alone. With funding from the Clinton Climate Initiative – a project set up by former US President Bill Clinton's Foundation – Livingstone has brought together leaders from 40 of the world's largest cities to reduce urban greenhouse gas emissions. The so-called C40 Mayors launched a global programme to make commercial and public sector buildings more energy efficient. Other cities taking part include Delhi, Karachi, São Paulo, New York, Mexico City and Seoul.

... and Cardiff

Understanding where a city's emissions come from is essential for planning how to cut them, according to Dr Andrea Collins at the BRASS Research Centre in Cardiff. She worked with Cardiff's City Council to calculate the city's environmental footprint – and found it was growing so quickly that even stabilising it would be a challenge.

Further studies have looked at how Cardiff could reduce the impact of events held in the city. The 2004 FA Cup between Manchester United and Millwall, hosted at Cardiff's Millennium Stadium, generated an environmental footprint of 3,051 global hectares (see Chapter 3) – with emissions from travel, take-away food and bottled drinks. 'You know where the footprint is going to come from for that event, so you can look at waste or transport, and see how they are going to be affected,' Collins explains. 'It should mean that people take a more joined-up approach.'

home is where the heat is

Homes account for around a quarter of all UK carbon emissions. Collectively we can do a lot to cut emissions from homes – while making them better to live in and cheaper to run.

To date UK Government policy has focused on emissions from new houses, setting a target for all new homes to be zero carbon by 2016. But of the homes we will be living in by the middle of this century, up to three quarters have already been built – it is the old housing stock that requires most attention.

Research in 2007 by the Environmental Change Institute at Oxford University said UK homes could slash their emissions by 80 per cent by 2050. The research sets out a programme that would be rolled out street by street, village by village, city by city until every home is fully insulated and has small-scale renewables. It will require the Government to use a mix of tough regulation, high minimum standards and generous subsidies and tax breaks. In a nutshell here's how it would work:

- Give power and funding to local authorities: they can identify people who need financial help most urgently to upgrade their homes.
- Make it illegal to sell the most energy-inefficient homes: do this through minimum standards and by ensuring every home has an energy performance certificate.
- Create legally binding targets at national and local authority level for reducing emissions from housing.
- Roll out community-wide combined heat and power plants to help make the most of the energy we use.
- Set tough minimum standards for lighting and appliances to ensure only the most efficient and low-energy ones are used.
- Ensure that people can sell their excess electricity to the grid for a premium price – making micro-generation more appealing financially.
- Provide financial support such as tax rebates and large grants so that everyone can afford to insulate properly and generate their own power.
- Install electricity and gas monitors and meters in every home to help people save energy.

The research, commissioned by Friends of the Earth and The Cooperative Bank, says that if these policies are implemented in full, by 2050 every household could have good insulation and solar power. We would be warmer, have more hot water and could even run more appliances than now. No household would spend more than 10 per cent of its income on energy. Fuel poverty will be history. It is an exciting vision that is simply streets ahead of where we are now.

grassroots action

Government and local authority leadership will have to play a massive role in shifting society towards a lower carbon lifestyle. But a sense of urgency and personal responsibility is driving many people to find ways to cut carbon at home and in their neighbourhoods.

case study
parish news

Tony Wainwright lives in rural Cornwall, a long way from Whitehall strategies.

Involved in a county-wide network of people wanting to take action on climate change, Tony decided to see if people in his local community could be motivated to turn St Endellion – a cluster of hamlets and fishing communities in North Cornwall – into a climate friendly parish. The aim: to cut emissions by 5 per cent every year.

'I thought maybe a parish would be the right level for doing something,' Tony explains. 'I was thinking about how villages work. People talk to each other, they can say "This seems to be a great idea, let's do it."' Tony approached the parish council, who supported the idea and set up a committee in September 2006 to take it forward.

The first task was to work out how big St Endellion's carbon footprint was. With the help of students from the local college, they looked at the impact of domestic energy use and farming and at the potential for a local offsetting scheme. 'We thought that if we could set up a local scheme for offsetting emissions, we could have a fund and use the money to put energy efficiency measures in public buildings like the village hall and the church,' Tony says.

Farmers have been encouraged to get involved in plans for a local food economy – one runs a shop stocked with local produce and is keen to develop the idea. Public transport is limited, so plans are being hatched for lift-sharing and a park-and-ride scheme. More eco-tourism and green plans for businesses are in the pipeline. The local pub, the Crow's Nest, carried out an energy audit and is aiming to become a climate friendly pub. The project has also organised a screening of Al Gore's *An Inconvenient Truth* to raise awareness of climate change, as well as a local food tasting day and a green fair.

Not everyone has wanted to get involved, however, and Tony says he has discovered things about his community. 'You really are dealing with the relationships that people have with each other,' he says. But he hopes St Endellion's experiences will help other parish councils and is producing a guide on what they have done.

case study
Ashton Hayes

In 2005, during a year off from running his own business, Garry Charnock went to a public debate about climate change.

'I came away thinking what could I do?' he says. 'So I rang up the Energy Saving Trust and I asked if they could put me in touch with any villages who've tried to do this, and they got very excited.

'I went to see a few friends at the pub quiz. I got such a good response I decided to go to the parish council and appeal to them to consider becoming the first carbon-neutral village in England, and that if they did it I would promise to run the project.'

When the Energy Saving Trust found out how serious Garry was, they offered to come along to the parish council meeting to back him up. The added presence of the University of Chester, and a larger-than-average turnout from the village, made for a packed meeting. And the

TREE CHEERS: Ashton Hayes won the Community Initiative prize for its Going Carbon Neutral project at the Energy Institute 2007 awards.

verdict? 'It wasn't unanimous but they voted to do it.'

Ashton Hayes (population 1,000) has not looked back in its quest to become England's first carbon-neutral village.

The launch meeting took place in January 2006 – attended by 400 people. 'The key thing was to hold it at the school. Before the launch we got the kids to do a project on how they would design clean vehicles for the future, so they all went home and discussed with their parents and they came up with all these drawings and it meant we had an exhibition in the school for the night, which drew the parents there as well. When the journalists came, they put the kids in front of this display. And that was the learning we've had all along, that carbon dioxide is invisible and you have to think of visual ways of engaging people with it.'

The first year's measurements showed Ashton had reduced its carbon footprint by 20 per cent, exceeding all expectations. Going Carbon Neutral has now attracted global attention. A short film on the project includes a clip of publican Barry Cooney saying, 'People aren't talking about Man United or Liverpool in here now, they're talking about climate change.' His pub, The Golden Lion, is an active part of the campaign – and its own efforts to reduce its carbon emissions have been backed by their brewery. People in a small town called Castelmaine (near Melbourne, Australia) were so inspired by Ashton's campaign that they want to do the same thing. Castelmaine and Ashton Hayes are now twinned and update each other on progress.

A key factor in Ashton Hayes' success to date seems to be the strength of the community, Garry says. 'We've now got about forty volunteers – all those people standing up are extremely busy people and decided that the buck has to stop with them. And it's not just the residents, it's local businesses and councillors who've all joined with us. Working together has made things happen quickly, has created no barriers, and has actually been fun.'

case study
changing perceptions in Leicester

The Muslim Khatri Association was set up in 1977 to cater for Leicester's Khatri community. Its centre, in the inner city, has become a model eco building, with high standards of insulation, low-energy lighting, environment-friendly flooring and paints, and solar PV panels on the roof.

Yahya Thadha, who manages the project, explains that the green features emerged from a consultation with the local community when the building was redeveloped in 2002. They had watched films about a local eco-friendly demonstration house together and suggested the centre should have solar energy and other environmental features.

Getting planning permission was a major hurdle as the centre was in a conservation area, but local residents and community members lobbied the council and got the conservation status removed.

Traditionally, jobs sorting rubbish have been associated with the lowest caste in society, making some people reluctant to recycle. 'It's changing those sort of perceptions, and standing up with the community, but actually making them think it's all good for the future,' Yahya says.

The refurbished centre has gone from strength to strength, running education classes for adults and children, holiday activities, a day-care centre for older people and an IT centre. Electricity from the solar panels is being sold back to the grid, and the centre is looking at installing a wind turbine or ground-source heat pump.

'We ended up with a model building – what people actually wanted,' Yahya says. 'It was a turnaround, from it only being used 10 per cent [of the time], now we have almost 85 per cent occupancy, and at peak times we're having to turn people away.'

CARING COMMUNITY:

Redevelopment of Leicester's Muslim Khatri Association centre as an eco-building shows grassroots action can inspire change locally.

case study

making the transition to a fossil-fuel free world

Rob Hopkins is a man on a mission. He wants the world to face up to the question of what will happen when the oil runs out. His determination has seeded a network of Transition Towns, communities across the UK that are finding ways to grow their own food, invest in renewable energy and develop local economies.

The idea of making the transition to a carbon-free economy was born out of research into the world's oil supplies and the idea that we may have already passed the peak supply of oil – combined with a response to the threat posed by climate change.

Hopkins, who was teaching permaculture in Kinsale, Ireland at the time, asked his students to work on how their local community could make the shift from fossil fuels. The result was an Energy Descent Action Plan, describing how Kinsale could reduce its dependency on coal, oil and gas by 2021. 'If we are looking at life with a quarter or a fifth of the oil available now,' Rob says, 'you need to create a dynamic vision of where we are going.'

The Transition movement draws inspiration from places that have already undergone a fundamental energy shift, such as Cuba, which lost half its oil supply when the Soviet Union collapsed. At first the impact was disastrous for Cuba's economy and people, who went short of food, drugs and transport fuel. But the country responded with a community-led programme of urban gardening, providing local organic food, a reliance on cycling and shared transport schemes and government-led investment in renewable energy. The movement also draws on the memories of older generations who grew up in a world with less oil and who remember how their local communities used to function.

Ben Brangwyn, co-founder of the Transition Network, says we don't have to go backwards. 'It could be the 1930s with better gadgets, far superior medicine and dentistry and much greater connectedness thanks to the web,' he says. 'There'll be far better schooling and less polluting industries.'

Totnes, in Devon, began its move to a Transition Town in 2005. A core group of volunteers spent the first year spreading the word among locals about the need to reduce fossil-fuel dependence. Rob Hopkins brought volunteers together from across the community to discuss what they thought the town should do, before 'unleashing' the project in September 2006. 'It's about asking the right questions,' says Rob. 'This whole thing is like a large research project. Maybe it will work – we will find out.'

As a result of the experiment, Totnes has set itself a solar power challenge, encouraging 50 homes to invest in solar hot water. It has established its own currency – the Totnes pound – with bank notes that can only be used at local stores. The town council supports the initiative and has helped plant nut trees in open spaces around the town to provide a source of local protein. A Totnes Energy Saving Company has being set up.

The model has spread: communities from Lampeter to Lewes are developing transition plans, with the Transition Network on hand to advise.

Worldwide, a network of more than 100 communities is finding ways to rely on local resources in preparation for life with less oil.

personal power

Whether it's the way we get to work, heat the house, use gadgets, or what we choose to buy, as individuals we clearly have an effect on total climate-changing emissions. But how much control do we each have over our emissions?

According to the government individual behaviour accounts for 40 per cent of the UK's greenhouse gas emissions; research by the Centre for Alternative Technology (CAT) suggests individuals can only directly affect about a third of their per capita emissions (in other words, the emissions we're each responsible for if you divide the national total across the population). The CAT research suggests about half of our per capita emissions can be traced back to the goods we buy – and we can reduce some of this through the choices we make. But about 15 per cent of our per capita emissions come from the social infrastructure that we all rely on, such as government buildings and services, but over which we as individuals have little direct control.

Leave aside for the moment the 60 per cent of total emissions we do not directly control, can our behaviour as consumers really have much impact? Can we buy our way out of climate change?

We don't always have the time, money or information to make choices with the lowest environmental impact. Leaving the car at home may not be realistic if you live in an area where the bus comes only once a day. Upgrading to an energy-efficient fridge will not be a priority for many families: there may be more pressing calls on the family purse.

Although we can each do things to curb the energy we use at home, it is government policies that determine the cost and availability of the other energy we use. That's why government action is essential if enough people are to shift to a low-carbon lifestyle. Action from business is also vital; but many people believe there is no substitute for the government setting clear standards for reducing emissions levels across all sectors, making it easier for the rest of us to do our bit. Our councillors, MPs and members of the European Parliament could support higher standards to achieve the carbon cuts that the science says we need.

Personal carbon emissions

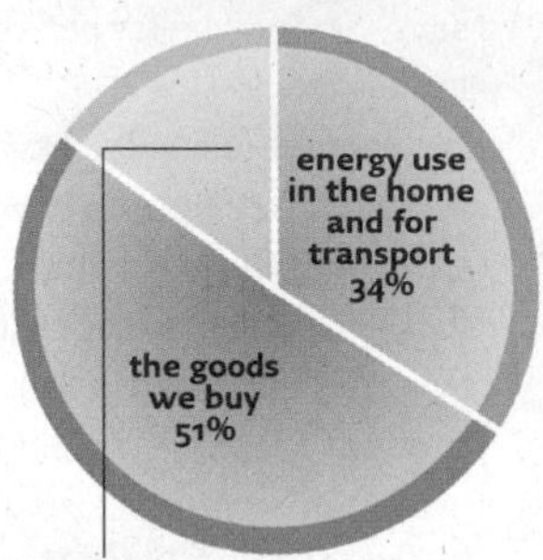

setting personal limits

One suggestion for cutting our collective carbon footprint is the individual carbon ration or carbon allowance. This takes the idea of putting a limit on the carbon we can all emit together by allocating a carbon share to individuals.

Personal carbon allowances, in theory, give individuals the chance to decide how to use their share – choosing, say, between making a long-haul flight for a holiday or driving their car to work every day. Anyone not using their full carbon quota would be allowed to sell allowances; and anyone failing to curb their carbon use would need to buy carbon credits. This could provide a clear financial incentive to cut personal carbon use and stimulate a market for low-carbon goods and services.

The Royal Society for the Encouragement of the Arts, Manufactures and Commerce (RSA) looked at how such a scheme might work. It suggests quotas could be set in line with a UK national carbon budget (established under the Climate Change Law). Individuals could open carbon accounts with a weekly or monthly allowance that they could spend or sell; the accounts could be managed through a banking system whereby individuals would get either carbon credit cards, or pre-paid carbon on a swipe card. Labels on products would show the carbon price of goods; the carbon price could also be included on domestic energy use, public transport, aviation, petrol and diesel. A voluntary scheme of this type, suggests the

RSA, could be working by 2010; but it acknowledges that a voluntary approach is unlikely to lead to deep enough cuts in carbon, and suggests that once up and running, it could be made compulsory.

Making such a system fair is recognised as a problem. Different groups in society rely on carbon in different ways – a farmer living in an isolated rural part of Scotland arguably needs more carbon for travel than a city-dwelling office worker with good access to public transport.

There is also a risk that the least well-off in society would benefit least from such a system as they would not be able to afford to supplement their carbon quota. Almost a third of the UK's least well-off families use more energy than the national average because their houses are so badly insulated, which means they would have to pay for the extra burden of using more energy for heating, without necessarily having the means to improve their insulation. People on higher incomes would be able to buy up surplus carbon allowances, giving them easy access to high-carbon activities such as long-haul flights.

Research suggests that a quota covering energy use in the home and transport could balance out: people in poor housing would be less likely to use their carbon travel allowance and so would be able to supplement their heating needs.

One option would be to use carbon trading to tackle only certain areas of energy use – for example, domestic heating and transport fuel. Curbing emissions from the manufacture and production of the goods could be done by targeting manufacturers and retailers, and would then be reflected in the price.

Some people have embraced the idea of personal carbon allowances, setting up local schemes through the Carbon Rationing Action Group network (CRAG). CRAGgers agree a personal carbon budget within their group and set up penalty or incentive schemes to encourage those involved to stay on track.

greening the National Trust

In 2007 The National Trust put action to tackle climate change at the heart of its future plans, urging its 3.5 million members to join the effort.

The Trust has been working to cut the carbon in its own backyard for some time. Its head office in Swindon was built to high environmental standards, maximising daylight and using natural ventilation and solar photo voltaic panels to generate electricity.

Away from head office, Gibson Mill in West Yorkshire is the Trust's flagship eco property following an 18-month redevelopment to improve its eco-credentials. Visitors to the 200-year-old cotton mill at Hardcastle Crags must arrive on foot; and when they do they find displays about renewable energy and sustainability alongside the mill's history.

Guy Laurie, Property Manager for Gibson Mill, says, 'If you go to a stately home, all this is tucked away, but here we're very brazen. Our message is about renewables, sustainability, environmental impact, carbon footprint etc so why hide it? It's the same with the photovoltaics – no need to hide them – they're there.'

Although some within the Trust wanted a more conventional refurbishment, including connecting the Grade 2 listed building to mains electricity, others argued that the mill had been built to make use of natural resources and there was no reason why it couldn't continue that tradition but bring it up to date. This meant convincing the Trust's architectural panel and a reluctant local council of the need to install solar tiles on the roof, to add to the electricity generated by hydro turbines on the river.

A wood-burning stove provides hot water, with fuel harvested from the surrounding woodlands (right). The solar tiles and hydro together can generate more electricity than the building uses. New appliances are monitored for energy use to ensure the system is not overstretched.

'We've been at this now for over two years,' Laurie says. 'We've never failed to open to the public. We've always had the electricity and the hot water.'

consumer power – does it matter?

While we need government action to switch society quickly to a low-carbon path, the choices we all make as consumers can have some influence, both in persuading business to change the way they produce the goods we buy, and in saying to government: 'We're ready for change – now you give us the policies that will make it happen.'

Ethical consumerism has boomed recently. From bank accounts to washing-up liquid, there has been an explosion in products and services claiming to be environmentally sound. Many businesses claim they are taking steps to tackle climate change; many claim their products are green. But how is the average shopper to know the real impact of the products they buy? Even if you want to do the right thing, it's not always easy figuring out what the right thing is.

Research in 2007 for Consumers International – a group representing consumers' interests – found shoppers could make a difference by opting for environmentally friendly products, but that more help was needed to guide the decisions they made. The research found companies had to be more tightly regulated to ensure high environmental standards; and consumers need clearer labels to show what they are buying. Clearly there is a vital role for government in regulating product standards for their environmental impact just as it does for health and safety.

Business activity is responsible for an estimated 40 per cent of the UK's carbon emissions. Government initiatives such as the Climate Change Levy and Emissions Trading Scheme attempt to drive the market towards cutting carbon but consumer pressure can be a powerful driver as well. Public concern about climate change, reflected in consumer demand as well as political action, has helped push the issue up the political agenda and sent a message to government that this is something the public wants them to act on.

Companies are responding to climate change in all sorts of ways, from taking it seriously and doing something about it to simply seeing a new marketing opportunity and cynically jumping on the bandwagon.

literary footprints

Independent of government, national or local – there are opportunities to curb the footprint of all kinds of events. The Hay on Wye Festival – an annual literature festival once described by former US President Bill Clinton as 'the Woodstock of the mind' – which attracts around 80,000 visitors, set itself a target of cutting its carbon footprint by 20 per cent over three years.

In 2006 the festival team launched a scheme to make the event more sustainable, working with other festival organisers to find the best ways of doing things. Sustainability Officer Andy Fryers explains: 'Hay Festival has grown from strength to strength; however with that growth we have become increasingly aware of the impact that we, and our audience, have on our environment.' How the festival is managed can make a real difference, and the team has looked at everything from stage lighting to bottled water. Making the most of natural daylight can reduce electricity consumption, for example, while providing water standpipes means visitors can refill water bottles. Local suppliers are used where possible, reducing the travel footprint of setting up the festival – and the 80,000 visitors are encouraged to share cars or use public transport.

'Within the festival team, everyone was all for it so it was reasonably straightforward to start the ball rolling and begin making the major change,' Andy says. 'What is surprising is the ease with which many people are willing to change when they see (a) what is possible and (b) that others are doing it.'

At the 2007 Hay Festival environmental campaigner George Monbiot suggested that in order to tackle carbon emissions such festivals should take advantage of new technology to become virtual events. Later that year Canadian author Margaret Atwood attended the Cheltenham Literature Festival by video link and signed books for fans using a new remote device called the Longpen.

the climate change levy

The Climate Change Levy was introduced in 2001 to encourage business, industry and the public sector to use energy more efficiently. It works by taxing business for the energy it uses – with renewable energy exempt.

Is it effective? Experts claim the levy could be leading to savings of 3.5 million tonnes of carbon a year, with a reduction in energy demand across the economy of 2.9 per cent a year. The Levy has also helped reduce business energy costs.

it's the government, stupid

'As the science of climate change is widely accepted, public attitudes will make it increasingly difficult for political leaders around the world to downplay the importance of serious action to respond to the challenge.' *(Stern Review)*

Government action, whether through law or policy commitments, is crucial if efforts to reduce our carbon footprint are to be effective across society. Government can direct individual and business behaviour; and it can ensure that the best choice for the climate is the easy, reasonable and most desirable one for most to make. If government uses its power well – over business, taxation, economic development – then individuals and business will know that their actions are contributing to a collective effort.

So what tools is the UK Government using, and how well are they working?

budgeting our carbon

Governments manage their money. They need to manage their greenhouse gas emissions too. Bringing our global emissions under control will mean setting a carbon dioxide budget – that way we can try to manage emissions across the world economy.

At a national level, governments can allocate emissions to different economic players. Each must stick within its carbon ration. Go above the carbon budget and there's a debt that must be repaid in order to avoid abrupt climate change.

In theory carbon budgets can work for individuals too. If you have 1 tonne of carbon to burn every year you can budget to spend that carbon as you choose – on heating, lighting and playing your stereo, on eating cheese and onion crisps or flying to New York. But if there just aren't the low-carbon goods and services on offer, individuals will have a hard time making a big difference. Government and business can facilitate the wholesale switch to a low-carbon lifestyle.

As far as the electricity sector goes – which is where government has concentrated most of its effort – companies in the UK operate in an open market, selling electricity to make a profit. The cost of the electricity is determined in part by the price of the fuel used for generating (mainly coal and gas). But if shrinking our carbon footprint means using (and therefore selling) less energy, can this be achieved by companies that depend on sales for their profits?

Measures such as the Climate Change Levy (a flat tax on emissions) and the Emissions Trading Scheme are intended to encourage electricity generators to reduce emissions by adding the cost of carbon to the cost of generation. But the low price of carbon has meant this has not succeeded.

Another Government initiative, the Energy Efficiency Commitment, requires gas and electricity companies to promote and install energy-efficiency measures in customers' homes. In some areas Energy Service Companies have been set up, often in partnership with local authorities, to deliver energy services (including efficiency measures) rather than simply sell energy.

The UK market has been praised for its high level of efficiency, although consumers have faced steep price rises when generation costs have risen without seeing price cuts when generating costs go down. What's more, the failure to reflect the cost of carbon pollution in the electricity price means business users and consumers are encouraged to use energy without paying the full bill for the pollution caused.

UK policy instruments - how are they measuring up?

the renewables obligation

The main policy tool to encourage green energy in Britain is the Renewables Obligation. This requires electricity supply companies to be getting an increasing proportion of their power from renewable sources, so that by 2015/16 renewables will represent 15.4 per cent of the power they supply.

How does it work? The companies making the electricity get a certificate for every megawatt hour of renewable energy they produce; the certificate passes to an electricity supply company when it purchases renewable energy. Any supply company that does not buy its quota of green electricity must buy certificates from another company which has exceeded its quota, or pay a fine. Consumers don't notice a change in their power supply, but a small percentage of their power is clean and green. Wind power, solar, small-scale hydropower, wave power, geothermal energy, energy crops, gas from landfill and sewage, tidal and biomass all qualify as renewables under the Obligation, as do some forms of energy from waste.

Has it worked? The Renewables Obligation has been described as a way of encouraging the 'least expensive' large-scale renewable energy sources. It has led to an increase in the amount of electricity generated from wind power, with renewable supplies now at 4.6 per cent - still below the 2006/07 target. However, it has not stimulated more use of microgeneration by households, communities or businesses - which has the potential to deliver significant cuts in carbon emissions. Uncertainty over the value of certificates in the medium and long-term also deters some investors.

Friends of the Earth verdict: with modification to make pricing more predictable to encourage greater investment the Renewables Obligation should continue to be the main - but not the only - way of delivering large-scale renewable energy capacity.

emissions trading

One powerful way of making sure the UK stays within its carbon budget is to make electricity generating companies and other companies that use large amounts of fossil fuels pay for the carbon dioxide they emit. If generating companies have to pay a high price for carbon, they are more likely to invest in efficient technologies and low-carbon renewable supplies as using gas and coal becomes more expensive. If the cost of carbon is low, then coal and gas - and even dirtier fuels - become more attractive financially.

Taxes and emissions trading are the main two methods of affecting price to cut emissions. All EU countries use emissions trading through a scheme set up in 2005.

How does this work? The EU sets a cap on total emissions from the types of installations covered by the scheme, such as power stations and big industrial sites. Each has an emissions allowance for a set period. Trading of these allowances happens, for example when some companies find cheap ways to cut emissions and can sell allowances to firms that need more because, for example, they are producing more goods. In theory the scheme delivers a set cut in emissions (through the cap) in the cheapest way (through trading).

Friends of the Earth verdict: The emissions cap in the first phase of the EU scheme was too high to ensure companies' investment decisions considered carbon. The price of carbon collapsed. The second phase will have a tighter cap but whether it will be low enough to make a big enough difference to the price of carbon remains to be seen.

kick-starting renewables - feed-in tariffs

If you're a home-owner or small business you're more likely to splash out on solar panels if you know you'll be able to sell electricity to the grid at a premium rate.

Many Governments use a financial incentive called a renewable energy feed-in tariff (REFIT) to support the development of green energy. This requires utilities to buy a certain amount of renewable electricity at a premium price. A feed-in tariff provides a guaranteed market for suppliers of renewable energy, making it easier to secure investment.

So do feed-in tariffs work? More than 40 countries, from Sri Lanka to France, the United States and Germany, have used them. In many European countries they have proved highly successful in encouraging the installation of a wider range of renewables technologies, including microgeneration. In Germany renewables supplied more than a tenth of the electricity market in 2006, with some 19,000 onshore wind farms. The sector employs more than 200,000 people.

Although critics argue that feed-in tariffs leave renewables suppliers vulnerable when the tariffs dry up, studies suggest they are as good at creating competition as the UK's current preferred option, the Renewables Obligation, and provide more support for less developed technologies.

Friends of the Earth verdict: A feed-in tariff in the UK would revolutionise the market, providing long-term security to homeowners and small and medium-sized businesses wishing to install microgeneration.

pricing carbon

The UK government is committed to introducing a carbon price into all its decisions. But Friends of the Earth believes the price it uses (£25.5 per tonne of carbon dioxide equivalent in 2007 prices) is too low to drive strong policies to tackle climate change, or to reflect the damage climate change is causing. This is for three reasons.

First, the price does not value many impacts, such as the costs of damage to nature or people's homes. Second, it is based on old science - more recent science shows climate change damage is likely to be worse than previously predicted. Third, the government makes unreasonable assumptions about the future. Because carbon dioxide stays in the atmosphere many decades, the damage done by a tonne emitted now depends on what sort of future we have. If we stop climate change, the cost now is quite low. And the government assumes that the worst climate change will be stopped, so it uses a low carbon-damage cost. But using a low value in deciding policy means climate change policies will not be strong – so the worst of climate change won't in fact be stopped. This is as if before a war, ministers said: 'We've looked into the future, and decided we're going to win. So we've factored that result into our plans and decided we don't need to spend any money on bullets or planes.'

Friends of the Earth verdict: The government must make reasonable assumptions about the future and set a much higher carbon price to stop the worst of climate change.

case study

doing business differently

The Co-operative Bank introduced an Ethical Policy for its investments back in 1998. The bank does not invest in any business whose core activity contributes to climate change through the extraction of fossil fuels or through deforestation. That means that unlike the other UK high street banks, it does not invest in any oil or gas company – or in any of the companies that provide related services – refusing an estimated millions of pounds of income per annum because of its ecological impact. 'It does have a big impact on the business,' Ryan Brightwell from The Co-op Bank says. 'But a third of our customers are with us because of our ethical policy. It does drive business as well.' (A survey in 2005 found that 34 per cent of the bank's profits came from customers choosing the Co-op because of its outlook on ethical issues.)

The Co-op Group, which includes insurance services and the Co-op shops, is investing in energy-efficiency measures with the aim of cutting carbon dioxide emissions by 25 per cent by 2012. It has also committed to renewable energy supplies, with its own wind farm near Cambridge, and plans to build a six-turbine wind farm in Lincolnshire in partnership with the electricity company Ecotricity. The group's Solar Tower in Manchester is one of the world's largest solar PV energy projects. Nineteen micro-turbines on a separate office building generate more power. The group is looking at reducing energy use across its activities, with a recycling centre in the office and new initiatives to cut the amount of packaging used for products sold in Co-op stores.

power of the law

In November 2006 the UK government announced that it would be introducing new legislation in the form of a Climate Change Bill to help the UK shift to a low-carbon economy. The bill, scheduled to become law in 2008, will set legally binding targets for emissions reductions. The Government will have to report back to Parliament on its performance, and will be held accountable by an independent Climate Change Committee.

The UK legislation, the first of its kind, is in large part the result of a campaign by Friends of the Earth, which generated hundreds of thousands of messages of support from members of the public, substantial media coverage and strong backing across political parties.

A number of other governments have set legally-binding targets for emission reductions, or announced their intention to make major cuts:

- In Germany, Angela Merkel has been pressing for a 40 per cent cut by 2020 and an 80 per cent cut by 2050.
- Costa Rica aims to be the world's first carbon-neutral country, aiming to cut or offset all emissions by 2030. Already generating most of its power from hydro-electric, wind and geothermal sources, the Central American state plans to tackle emissions from farming and transport, as well as using a scheme, set up in 1997, to pay farmers and landowners for planting trees.
- Norway is going for carbon-neutral status, cutting emissions by 30 per cent by 2020, and aiming to cancel out carbon dioxide emissions completely by 2050.
- Neighbouring Sweden has gone even further, saying it will wean itself off oil by 2020. The country has already made major steps to convert domestic energy to geothermal power or waste heat and is looking at using more biofuels for transport.

- Iceland wants to use hydrogen – generated using renewable electricity – to power all boats and cars by 2050.
- Although the US Federal Government has been slow to act, individual states are moving ahead. California has passed legislation to cut greenhouse gas emissions to 2000 levels by 2020 and by 80 per cent of 1990 levels by 2050. Separate legislation requires a cut in greenhouse gas emissions for vehicles.
- India's Integrated Energy Policy aims to reduce by a third the Indian's economy's greenhouse-gas intensity – that is, carbon output measured in relation to economic output.

Just as one individual's action to tackle climate change won't be enough to reduce the UK's carbon footprint, one country acting alone cannot save the planet. Climate change is a global problem needing a global solution. But how can individual countries work together to make sure emissions savings add up?

Europe has attempted to show leadership. It has set a Europe-wide target to go beyond the Kyoto Protocol commitments and cut emissions by 20 per cent of 1990 levels by 2020. If others follow suit, European countries have promised to cut emissions by 30 per cent. Germany has said it will commit to bigger cuts still – 40 per cent by 2020 – if Europe adopts the 30 per cent reduction target.

“ Clean air, clean seas and ecological balance are some of humankind's most fundamental shared goods. We cannot leave it to market forces and free competition to safeguard them.

Jens Stoltenberg, Prime Minister, Norway (20 April 2007)

policies to tackle climate change

Experts at the IPCC have looked at the need for government action to provide a framework for reducing emissions. Their recommendations include:

Regulations and standards: New laws and regulations can set standards for efficiency, for example requiring car manufacturers to make cars that are less polluting.

Taxes and charges: Financial measures can be used to make it more expensive to pollute by putting a price on carbon. But the experts warn that there are no guarantees as to how much carbon such approaches will save: industry will pay the tax and go on emitting if they profit by doing so.

WHERE THE POWER LIES:
Policies to combat climate change have to come from governments. And that happens when people tell politicians to get on with it.

> **“ It will be far easier to take policy forward in one country if other countries move together. We have the time and the knowledge to act. But only if we act internationally, strongly and urgently.**
>
> *Sir Nicholas Stern*

Tradable permits: Used in the European Union's Emissions Trading Scheme, tradable permits allow businesses that have reduced emissions to make a profit at the expense of those whose emissions are high. But permits are worthless if too many are issued.

Financial incentives (subsidies and tax credits): These can be used to encourage certain behaviours; for example, a subsidy on energy-saving light bulbs could boost sales. Such incentives can help new technologies compete.

Voluntary agreements between industry and governments: This is where governments ask business to reduce emissions, but do not enforce the request. It can be a useful way of raising awareness but it has not delivered significant emissions cuts.

Awareness-raising campaigns: Awareness-raising campaigns can help people make informed choices and possibly contribute to behavioural change. It's difficult to identify their real impact on emissions.

Research and development into new technologies, such as renewable energy generation or energy-saving tools: Funding for research can stimulate new developments, reduce costs, and lead to emissions cuts.

action at all levels

Shifting to a low-carbon life means changing the way we plan our towns, organise our transport, how we behave at work and how we spend our holidays. It is about how we do business and how we run our economies. It is about how we manage our environment and look after the land.

The very things we need to do to reduce greenhouse gas emissions can actually improve the quality of life: warmer homes that are cheaper to heat, better air quality, and better health.

Business can benefit from cutting its energy use, getting more from the same amount of energy and from wasting less. New industries must develop to support a low-carbon life; there are opportunities for clean technologies, for managing waste and recovering valuable stuff from it, for better insulation and for energy advice.

Community action can be powerful, with people getting together to decide they want to change their neighbourhoods. Farmers' markets, lift shares and campaigns for better public transport and better recycling facilities all show what people getting together can do.

And while some choices are down to the individual, moving to a low-carbon life quickly enough is only possible with the right policies in place – from local authorities, national government, the European Union and at global level.

We can have a low-carbon future. There are lots of ways to cut our carbon footprint - and by adding up the changes, we can make the level of cuts required. These are choices that have to be made.

the UK on the big stage

The UK is an important player in international efforts to cut carbon dioxide emissions. As well as being one of the world's leading economies – and therefore in a position to shape the agenda – the UK has taken a leading role at international talks aimed at cutting emissions.

The UK also has a big debt to pay. It embraced the new technologies of the Industrial Revolution, and used its riches to build a large empire. Our carbon footprint goes back a long way. Alongside its historic responsibility for a big slice of global emissions, the UK has a role on the world stage; but if it is not doing everything it can to curb its own emissions, how can it demand cuts of big emitters such as the United States or China?

nine lives – tips for the UK government

The UK is likely to be the first in the world to set national legally-binding targets for cutting carbon dioxide emissions. So where next? How to turn this ground-breaking law into a real win for the planet? Here are Friends of the Earth's ideas:

1. More renewable energy, and fast:
A target for renewable, including combined heat and power, to supply at least 45 per cent of total electricity demand by 2020. A tenfold increase in research and development for technologies such as wave, solar, hydrogen fuel cells and advanced cables. Support for international renewable projects – such as offshore wind in the North Sea, a high-voltage European grid, and concentrated solar power in North Africa.

2. Clean up surface transport: Scrap motorway widening, and boost investment in public transport and other low-carbon forms of transport.

3. Curb aviation growth: Stop airport expansion plans and do more to make the cost of flying reflect the environmental damage it causes. Greater investment in rail could make this an attractive alternative to short-haul flights.

4. Make it easier and cheaper for people to go green: Use measures such as council tax and stamp duty rebates to make homes more energy efficient, and fuel-efficient cars cheaper than gas-guzzlers.

5. Roll out a nationwide eco-homes programme: Use tough minimum standards, generous financial support and better information to slash emissions from housing and wipe out fuel poverty.

6. Boost energy efficiency: Create a massive programme to ensure that energy is used more sensibly and efficiently.

7. Clean up coal: Invest in techniques for cutting emissions from coal-fired power stations.

8. Toughen planning policy: Oblige all new developments to get a fifth of their energy from renewables. Reward local planners that reduce carbon emissions in their area.

9. Set local government targets: Require all local governments to cut carbon emissions by 40 per cent by 2020 in their area. Woking has shown a 50 per cent cut is possible.

international action - Rio to Kyoto

The Rio Earth Summit in 1992 kick-started international action to tackle climate change. At this milestone meeting country leaders signed up to the United Nations Framework Convention on Climate Change (UNFCCC). Signed by 189 countries, the UNFCCC recognised the climate as common property for the whole world and set the long-term objective of stabilising greenhouse gas emissions to prevent 'dangerous' human interference in the climate system.

In 1997, under the convention, countries negotiated a set of legally-binding emissions cuts, called the Kyoto Protocol, which came into force in 2005. The Kyoto Protocol has been adopted by 141 countries. There were a few notable exceptions - the United States refused to join.

The current first phase of the protocol requires signed-up industrialised countries to reduce emissions by 5.2 per cent from 1990 levels by 2012. Each country reports on its emissions and will face penalties if it fails to comply. Under the protocol, rich countries and industries channel funds into some of the world's poorer countries for clean energy projects and can claim these funds represent emissions reductions. This so-called Clean Development Mechanism is meant to help clean up industry in some parts of the world, though critics argue that countries are claiming credits for things that should have happened anyway.

Under the Kyoto Protocol rich countries are also funding work to help the world's poorest countries adapt to the impacts of climate change.

The World Bank has estimated that the cost of protecting the world's poorest countries against climate change is between US $10 billion and US $ 40 billion, though this figure may be a gross underestimate. Talks on action after 2012 got off to a slow start in Bali in 2007 but countries are all agreed that they must reach a final agreement when they meet in Copenhagen in 2009.

Climate change is triggering a lot of high-level talk. But emissions are still rising. So is it all just a load of hot air? The answer is that the impact on emissions from the first phase of Kyoto is likely to be small; but the protocol has established the only international legal structure to tackle emissions and, even if its contribution to date is small, it may provide an important foundation for future efforts to address climate change.

If you want to make a personal contribution to saving greenhouse emissions, you could start at home. Take a look at what you can do to make a difference.

in the home

chapter 7

carbon in your home

Our homes are responsible for about a quarter of the UK's total annual carbon dioxide emissions, so they're a good place to start if you want to make a personal impact on greenhouse gas emissions. By buying carefully, and following simple energy tips, you can make a huge difference where it counts. This chapter looks at how you can cut your household carbon emissions in everything from eating and heating to insulation and lighting – and even tells you how to create the perfect eco-home.

Before going through these individually, here's a typical breakdown of carbon dioxide emissions from a gas-heated three-bedroomed semi-detached house.

Use	emissions (tonnes CO_2)	% of total
Space & water heating	4.4	70
Lighting & appliances	1.6	25
Cooking	0.3	5
Total:	**6.3**	**100**

Here are the areas you need to look at to cut down emissions in the home.

top tip

Much of this chapter is about things you can do to curb your emissions. But look out for waste: studies show that when we do bigger things like insulate the loft many of us simply allow the house to be hotter rather than turn the heating down to make the full savings.

INSULATION

It may not be sexy but insulation, whether of wall, roofs, or windows, is one of the best ways to shrink your carbon footprint – and it's cheap, too. While the latest building regulations mean new homes must have high levels of insulation, older houses – the types most of us live in – leak heat at an alarming rate. The more heat they leak the more energy they need to keep warm and the bigger the carbon footprint. There are around 17 million homes in the UK with cavity walls that can be easily insulated to slash heating bills but 11 million of those properties still have no cavity insulation. That means about 11 million tonnes of carbon dioxide or 1 tonne per house could easily be saved. But if you're not blessed with cavity walls, like the 7 million houses built with solid walls, don't just turn up the thermostat and hope for the best: whatever type of house you have there will be a form of insulation that can slash your gas bill, cut your carbon and make your home comfier all round.

A well-insulated building will have the instant effect of magnifying the savings created by any other energy-saving measures you adopt. If less heat escapes, your heating system won't need to work as hard, so reducing your bills, carbon emissions and demand for electricity and gas.

top tip

More than 50 energy efficiency advice centres nationwide can provide all sorts of advice on saving energy in your home. Freephone 0800 512 012.

insulation works – just ask an Eskimo (or 'blimey, it's as cold as a three-bed semi in here')

Snow isn't known for its warming qualities, but an igloo made of nothing but snow and ice can keep its occupants at a (reasonably) snug 0°C while outside it's as cold as -45°C. That's a very welcome 45°C difference, all thanks to a simple understanding of how a house should be insulated. The igloo manages to stay warm because it controls the air that comes in and out. Body heat melts the inside surfaces of the snow blocks and where the water refreezes the walls become sealed tight. The door passage always has one right angle in it to protect from the wind and the sleeping areas are raised up to make the most of warm air rising. It's this combination of thick walls and controlled air flow that means a small fire or candle and body heat will warm up, and keep the inside warm because the heat has nowhere to go. Now we're not suggesting you head to the freezer section for some ice cubes, but it just goes to show how important it is to insulate your home.

loft insulation

What is it?
A layer of insulating material between and over the ceiling joists in a roof space that acts as a blanket preventing heat from escaping as it rises. Loft insulation isn't new: 95 per cent of British homes will have some. Indeed, 2006 building regulations insist on a minimum insulating material thickness of 25 cm but many houses will have a thickness of only around 2.5 cm-7.5 cm. Topping up this level of insulation to at least 27 cm is recommended. If you use your attic as a room you can also insulate the roof joists to make the space more comfortable.
Can I do it myself?
Unless you opt for the 'blown' method (in which insulating materials are pumped into the roof space by machine) you can insulate your loft in a weekend. If this isn't an option for you then a professional installer can advise you on the best products for your home.
How much will it cost?
The government calculates that the average cost of bringing your loft up to current legislation would be £284, but the figure depends on the materials used, the size of loft and whether or not you're doing it yourself. Grants are available.
How much cash will I save?
Assuming you already have some insulation, and top it up to the recommended thickness, you can expect to save £50-£80 per year on your energy bills, so the insulation should pay for itself in four to five years. If you don't have any, the savings could be around £130 per year.
How much carbon will I save?
410 kg per year for a top-up – or a lot more (1.5 tonnes) if you have no loft insulation at all and install a layer 27 cm thick.
What materials should I use?
Basic rolls of fibreglass or mineral wool will provide superb insulation for very little cost (from just £25 per 10 square metre), but they require a lot more energy to produce compared to

many of the natural alternatives such as Excel Warmcel 100. This is a loose fibre made from 100 per cent recycled paper and is easy to install: simply pour into the loft space. It's quick, easy and cost-effective (£6 per m^2, Green Building Store). Thermafleece is another great choice, made from pure British sheep's wool, it's completely safe to handle and costs £112.52 for a $10m^2$ pack.

top tip

How much more will it cost to choose green?

The answer is, quite a lot more:
Thermafleece 100 mm thick – 10 m^2 costs £112.52 (Green Building Store)
Mineral wool insulation 100 mm thick – 14 m^2 costs £28.98 (Wickes)

cavity wall insulation

What is it?

Since the 1930s most housing has been built with an air cavity between the two layers of the outer walls and while this gap makes the house more energy efficient compared to older solid wall properties, massive improvements in efficiency can be made by drilling a small holes into the brick work and injecting the air gap with plastic foam or mineral wool.

Can I do it myself?

It's not a job for a DIY-er but a good contractor should be able to complete the work with very little disruption from the outside of the property in three to four hours, depending on the size of the house.

How much will it cost?

Prices vary, depending on the size of your house, but expect to pay around £260. Grants are available.

How much cash will I save?

According to the Energy Saving Trust (EST) you could see a reduction in your annual heating bill by one-third. The cost of the insulation would pay for itself in just two to five years.

How much carbon will I save?

This simple job could save the average home 1 tonne of carbon dioxide per year – that's 11 million tonnes of carbon dioxide that could easily and cheaply be saved, which makes it by far the simplest and most cost-effective way to cut your carbon emissions.

What materials should I use?
There is simply too much moisture in the wall for natural fibres to cope with. This means glass or mineral fibre, polystyrene, or polyurethane products are used.

fact

Between 2002 and 2005 around 800,000 households installed cavity wall insulation, saving nearly 400,000 tonnes of carbon dioxide, enough to fill Wembley Stadium 47 times.
(Energy Saving Trust)

internal wall insulation/dry lining

What is it?
Solid walls lose more heat than cavity walls so if your home was built before the 1930s you won't have a cavity to insulate. Instead you can increase the level of insulation by fixing insulated materials to the inside of the external walls of your house. This can be done either by fixing plasterboard-covered insulation direct to it, or by screwing battens to the wall, filling with insulation and covering with traditional plasterboard. Internal wall insulation will create a more even temperature in the room, eliminate condensation on the walls and slash your carbon emissions. Typically a thickness of 90 mm will give the best saving, but remember, it will decrease the overall size of a room, so if this is of concern thinner insulation is available. The maximum benefit comes in the rooms most used and heated.

Can I do it myself?
A competent DIY-er should be able to complete a basic internal insulation, but for larger rooms or very old properties calling in the professionals is recommended. It's worth including internal insulation if you are planning to redecorate because of the disruption it causes – you'll need to refit coving, skirting boards, plugs and sockets, and even radiators in some cases – but for not a great deal more you can have a newly painted highly insulated room all at the same time.

WARM INSIDE:

Consider the benefits of insulating internal walls. For more info contact Energy Saving Trust (0845 120 7799).

How much will it cost?

Costs vary greatly. As a rough guide £40 per square metre. For a whole house expect to pay £650-£2,000 depending on size. This price will be slightly higher if you choose environmentally friendly products. Payback can be as fast as four years but is usually not as quick as replacing an old boiler.

How much cash will I save?

This depends on the type of materials you use, as thermal insulating properties vary, but according to the Energy Saving Trust you can save between £110 and £490 a year, depending on the size and condition of your house.

How much carbon will I save?

The Energy Saving Trust estimates that a three-bed semi-

fact

If all the houses with unfilled cavity walls had them filled, the energy saved could heat 1.7 million homes each year.

(Energy Saving Trust)

insulate yourself

Remember, bricks and mortar don't mind being chilly, and will rarely complain – it's the inhabitants who tend to feel the cold. An extra blanket, thick woolly jumper or heavy curtains are all highly effective forms of insulation. Try one before you reach for the thermostat.

detached house could save a whopping 2.25 tonnes per year.

What materials should I use?

Thermafleece is made from pure British sheep's wool, takes a fraction of the energy to produce compared to glass fibre or mineral wool and is completely safe to handle. A $10m^2$ pack of 100 mm thick Thermafleeece costs £112.52 from The Green Building Store (www.thegreenbuildingstore.com). Alternatively, Pavatherm insulation board is made from wood pulp, is free from glue and can reduce heat loss significantly. Prices vary but expect to pay around £8 for a 60 x 1020 x 600 mm board or £102 for an external wall H 220 x W 350 cm.

external wall insulation

What is it?

Also known as cladding. The solid external walls of your building can be covered in highly insulating materials to reduce heat loss. It's increasingly popular as a way to improve the look and efficiency of old tower blocks and social housing in the public sector and as an alternative to internal wall insulation.

Can I do it myself?

External insulation often needs planning permission and should only be carried out by specialist companies.

How much will it cost?

It's not a cheap option – for a typical semi-detached house expect to pay £3,500-£5,500 or £45 to £65 per square metre. While the carbon and money savings are exceptional the payback period is far longer than with internal or cavity wall insulation. Lenders are starting to realise the value external cladding could add to some properties so both the Abbey and Nationwide now offer loans specifically for it.

How much cash will I save?

Around £290-350 per year.

How much carbon will I save?

According to the Insulated Render Cladding Association (INCA) each square metre of 50 mm insulation could save 1 tonne of carbon dioxide emissions over the life of a building. The Energy Saving Trust estimates savings of 2.4 tonnes of carbon dioxide per year.

What materials should I use?

Natural materials are available including the Pavatex range of natural fibreboard (www.sustainablebuildingsolutions.co.uk), which can be attached to your building before rendering. Lime renders (www.womersleys.co.uk) can also be used to create an environmentally friendly and very hard-wearing external seal to a building. For more information contact www.inca-ltd.org.uk; 01428 654011 or www.nationalinsulationassociation.org.uk; 01525 383313.

double-glazing

For every £100 you spend on heating about £10 is going straight out of the windows. So choosing the most insulating designs of window will make your house a nicer place to be, and your heating bills easier to pay.

What is it?

Double-glazing works by trapping air in a gap (ideally of 20 mm) between two panes of glass. It's this gap of air, rather than the thickness of the panes, than improves its efficiency. Available with wood, aluminium or uPVC frames, double-glazing will typically reduce heat loss by 60 per cent compared to single-glazing.

How much will it cost?

Replacing windows is an expensive way to cut carbon. For a three-bed semi expect to pay at least £3,000 for all new windows. Remember to choose wood products that are FSC-approved to ensure they come from sustainable well-managed forests.

How much cash will I save?

£40–£100 if you're upgrading from single to double glazing.

How much carbon will I save?

Savings of between 680 kg and 1 tonne of carbon dioxide per year compared to single-glazed windows are realistic.

What should I buy?

Look for the lowest possible U value (this refers to the thermal transmittance level – ie how much heat is lost through the panes). Since 2002 building regulations state that double-glazing must not have a U value greater than 2.0. Pilkington has an impressive selection of energy-efficient glass (www.pilkington.com).

Look for Window Energy Ratings (WER). These are a much more accurate indicator of the energy performance than U values because they take a range of factors into account, including solar heat gain. The British Fenestration Rating Council (BFRC) uses the same A-G rating (with A being the most efficient) as found on appliances, making it easy to see how efficient a particular model is. WERs are only voluntary as yet, but it's worth visiting the website (www.bfrc.org/) to check which manufacturers are involved and which energy-efficient windows are available.

Where possible, choose wooden frames such as the ECOplus system from Green Building Store (www.greenbuildingstore.co.uk) that use 100 per cent FSC-approved wood.

top tip

According to Giles Willson from the BFRC, 'the typical double-glazed window would be rated as E for energy efficiency, while some window manufacturers are producing A-rated designs that have a positive energy effect – actually letting in more heat than they lose.'

top tip

Consider retro-fitting double glazed panes to standard wooden frames. The Sash Window Workshop (www.sashwindow.com) has a service that can remove single panes from doors and wooden sashes and replace them with Bi-Glass double-glazing. It will cost around two-thirds as much as replacing the window frame, so you can keep all those original features and stay warm in the winter.

wood versus PVCu

	WOOD	PVCu
Cost (three bed semi)	£3000+	£3000+
Energy to manufacture	300 kWh/tonne	45,000 kWh/tonne
How polluting?	Wood from sustainable sources uses less energy to manufacture and won't add to carbon emissions and is completely recyclable.	Manufacture uses a range of harmful toxic chemicals and can't be recycled.
Lifespan	35+ years	25-35 years
Ease of repair	A joinery company can repair and refurbish all types of wooden frame.	Sealed plastic units are very difficult to conduct even minor repairs.
Maintenance	Can be sanded and repainted indefinitely.	Sold as 'maintenance free' PVCu windows have a reputation for getting dirty quickly and discolouring.

secondary glazing

What is it?
It is not always possible, or economical to fit new windows, especially if you live in a conservation area or a listed building where alterations are restricted. English Heritage estimates there are 44 million single-glazed windows in the UK. But you can greatly improve the efficiency of single-glazing by adding a second layer of glass or plastic to the inside of your existing window frame. Money and carbon savings are not as good for new double glazing but it's still worth considering.

How much will it cost?
Basic DIY window frames can cost as little as £300 per window and are available from your local hardware store. For professional made-to-order secondary glazing screens, expect to pay £600-£1,000, depending on size.

top tip
Draw your curtains at dusk. Sounds obvious, but thick curtains can stop heat from escaping through your windows.

case study
'it's oh so quiet'

'We live in a Victorian terrace with beautiful original windows, which are unfortunately single glazed, meaning our home was impossible to keep warm without turning the heating right up. We didn't want to lose the look of the windows but wanted to be warm, so we chose to fit secondary glazing. We paid £1,800 to insulate two huge sets of windows and it has totally, totally changed the warmth in our house. Even with thick curtains and windows stuffed with newspaper it used to be hard to keep the heat in. Now there is no sound from the street outside and the flat is markedly warmer. We keep the heating on for less time as a result and our bills have dropped too.'

Jo Finburgh, fundraiser, East London

How much cash will I save?
The National Energy Foundation estimates savings of £15-£25 per year and the EST suggest £20-£70 on your heating bill.
How much carbon will I save?
The amount of carbon you save will be dependent on the condition of the windows you are glazing. A very poor single- glazed frame that has secondary glazing could save up to 195 kg carbon dioxide, while a less inefficient design might only save 85 kg.
What should I buy?
The most basic form of secondary glazing is stretched cling film but this should only be considered as a temporary measure. There are many companies selling bespoke frames that fasten insulating frames onto the window with magnets, making it quick, simple and effective, with prices starting from around £90 per square metre. See www.magneglaze.co.uk

jargon busting

Look out for the U value when choosing any kind of insulation. It's a number used to describe how good a material is at insulating. Always look for the lowest U value. As a rough guide the U value of a single pane of glass is 5.4, standard double glazing is 2.6 and with high quality Low-E glass just 1.8.

draught proofing

Installing foam-sealant strips around doors and windows, fitting brush strips to the bottom of doors and around letter-boxes and using sealant along skirting boards and windows are just some of the quick, inexpensive ways to keep the heat in. Spending £50-£100 on draught reduction methods readily available (look under DIY, Hardware, in Yellow Pages for a local supplier) can save you £20 a year on heating and up to 140 kg of carbon dioxide. While the savings are not enormous you'll feel the benefits instantly. You can insulate a suspended wooden floor much the same as you can a loft with mineral wool, but without a basement it's a disruptive job as you'll need to raise the floorboards. That said, you can save 60 per cent of the heat loss in the room this way, so it's worth considering if you have to raise the floors to access pipe work etc in the future.

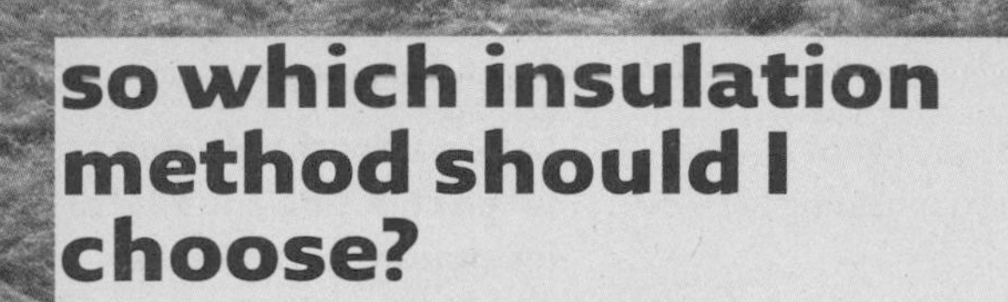

so which insulation method should I choose?

Not all properties will be able to benefit from all the different insulating methods. If you live in a ground floor flat, for example, you won't easily be able to fit loft insulation. Similarly, if you're in a Victorian or older house the chances are cavity wall insulation will be impossible. But the important thing to realise is that wherever you live you can do something to cut your carbon. It's unrealistic to think you can do them all – but there's always something you could do.

case study

getting better – room by room

Martin Normanton and his wife Jane have improved the energy efficiency of their Victorian semi in Walsall by more than 75 per cent over the past ten years.

Early retirement and a visit to the Hockerton Housing project (see p.296) inspired Martin – a campaigner with Walsall Friends of the Earth – to improve his own home. He describes the work as an on-going project. 'I'd been aware of climate change from the early nineties and that really made me decide I needed to do something about the house,' he says.

Installing a new condensing boiler and heating system made the biggest difference, Martin says, but internal wall insulation has also helped.

The house had been heated using electric storage heaters, so installing a well-designed central heating system, using a gas-fired condensing combi-boiler, thermostatic radiator valves and room thermostats made a huge difference cutting energy use by 60 per cent. Martin stresses that making good use of heating controls plays an important part in saving energy. Individual radiators can be turned down when rooms are not going to be in use, and the main thermostat is switched down when Martin and Jane are out for the day.

Internal wall insulation was added as each room was decorated, at a rough cost of £500 a room. Martin used polyurethane foam insulation sheets attached with battens to sheets of plasterboard, with expanding foam to seal all the joints. In the kitchen he saved a further 10 per cent by insulating below the level of the solid floor.

A plug-in electricity meter helped the couple identify which appliances were wasting most energy. Washing up by hand rather than using their dishwasher, they found, was more than seven times as efficient. Appliances on standby used a whopping 34 per cent of the couple's total electricity consumption. They have reduced their electricity consumption by half through changing their habits, a more efficient fridge-freezer and replacing old-fashioned lightbulbs.

where to cut the carbon

Here's a pathway to low-carbon nirvana – a list of practical steps in order of their cost effectiveness (payback). Take these steps one by one and you could slash your bills while curbing your carbon emissions: where you save money you'll also be saving the planet.

Based on government figures, the table has been amended in the light of Martin Normanton's hands-on experience (see p.253). It starts with the simplest measures such as turning down the thermostat and closing doors, which have the quickest payback. And it moves on to options that take longer to pay back, but have other advantages, like increasing the value of your home.

1 heating and hot water (typical pre-1988 three bedroom semi)

Energy saving measure	Bill £	Annual saving £	Cost of measure £	Payback (years)
Initial bill	**£780**			
Thermostat down 1°C	710	70	nil	immediate
Keep doors and windows closed	695	15	nil	immediate
Double insulate hot water tank	685	10	10	1.0
Cavity wall insulation	545	140	160	1.1
Draught strips (V seal)	520	25	20	0.8
Thermostatic radiator valves (TRVs)	480	40	140	3.5
Insulate hot water pipes	455	25	80	3.2
New condensing boiler	295	160	**R** 1,500	9.4
Top up loft insulation	270	25	**G** 160	6.4
Double glazing	220	50	**R** 3,300	66
Final electricity bill could be as low as	**£220**			

This typical house is a semi with gas central heating, built between 1930 and 1988. Changes to building regulations after 1988 required some cavity wall insulation. A pre-1930 mid terrace house will have similar heat requirements but does not have the option of cavity wall insulation. This example assumes that the usual basic measures of some loft insulation (1-4 inches), a cylinder insulation jacket and some draught proofing have already been done, shaving about £140 off the gas bill with payback periods of less than two years. If these items have not already been done, do them now. But for loft insulation go straight to ten inches, at a grant-aided price of about £160.

It also assumes that a 'room' thermostat is fitted. Some early central heating installations lack a room thermostat, and waste up to £120 a year of gas as a result, so if your system does not have a room thermostat, have one fitted straight away.

2 electricity (any home)

Energy saving measure	Bill £	Annual saving £	Cost of measure £	Payback (years)
Initial electricity bill	**£400**			
Switch off standbys, etc.	350	50 **1**	nil	immediate
Low-energy (CFL) bulbs	320	30 **2**	**R** 10	0.3
Replace old fridge freezer	270	50	**R** 225	4.5
Final electricity bill could be as low as	**£270**			

R: Replacement item. The payback periods assume that the item being replaced is in good condition; if it needs repair or replacement then the payback period is much better.
G: These items are usually grant aided, and the cost is after deducting typical grant. For grant details ring your local Energy Efficiency Advice Centre 0800 512 012.
1: £30-£80 potential saving.
2: This assumes that you already use low-energy bulbs in the most-used lights, so these figures are to use ten more low-energy bulbs in some less-used lights.

HEATING

Of the 6.3 tonnes of carbon dioxide emissions the average household is responsible for every year about two-thirds is for heating the space and water. So it makes sense to look into ways to cut the cost of heating and making the gas, oil or electricity you do use at home work as hard as possible. The main ways to do this are by insulation, controlling your heating better, thinking about the fuel you use to heat your homes, and installing a new, highly efficient boiler.

Since April 2005 anyone installing a new domestic gas boiler in the UK has been required by law (with very few exceptions) to choose a high-efficiency A- or B-rated condensing boiler. These boilers use up to 95 per cent of the energy you put into them to produce heat, compared to a more typical 50-60 per cent of older boilers. Put another way – for every £1 you spend on heating your house using an older boiler, at least 40 pence is going straight out the flue. A high-efficiency boiler also means your home will have a smaller carbon footprint and smaller bills because it needs considerably less gas or oil to maintain a comfortable temperature.

top tip

Electricity is probably the least efficient way of heating your home – so unless your place is small and really well insulated, consider installing central heating.

condensing boiler

What is it?

A central heating boiler burns gas (this is the most common type, though oil and biomass boilers are also available) to heat water. This hot water then heats all the radiators in your home and heats a hot water tank for use later (or instantly, if you have a combi boiler that doesn't require a storage tank.)

When the boiler is working, a proportion of the heat created by the gas escapes and is wasted. The difference between a modern condensing boiler and an older non-condensing design is simply that the modern variety converts the heat that would otherwise be wasted into useful energy – so you need less fuel for the same amount of useful heat. These new boilers look like any other design except for the addition of a small exhaust pipe (and gentle plume of vapour rising from it as the boiler works) that will need to be fitted on the outside of your house or flat.

How much will it cost?

In 2006 a condensing boiler would cost you £681– £1,045. Installation costs depend on the size of the house and condition of the heating system, but expect to pay between £1,500 and £3,000. They are considerably cheaper to run with a payback period of just four years in some cases. When thinking of carrying out any central heating work, always seek the advice of qualified CORGI (gas) or OFTEC (oil) installers.

How much cash will I save?

If your boiler is ten years old or more, chances are it will be losing as much as 44 per cent of the heat it produces; even a brand new one of conventional design will still be losing at least 20 per cent. A condensing boiler, on the other hand, can waste as little as 5 per cent. The result of this high efficiency could mean average annual savings of £100-£120 per house but that figure could rise to £240 if you have an extremely old (15+ years) boiler (EST saving assumptions 2007). That said, a boiler's efficiency greatly depends on how it's run, which thermostatic controls are used and what temperature it's set to run at.

How much carbon will I save?

The Centre for Alternative Technology calculated that if everyone in the UK had a condensing boiler our annual carbon dioxide emissions would be reduced by 17.5 million tonnes or 0.5 tonnes per household per year. The EST estimate is slightly higher at 830 kg CO2 per year per household.

Should I replace my existing boiler?

Thanks to the 2005 regulations the question is no longer should I buy a condensing boiler, but when? This also goes for your plumber or heating engineer, who should no longer be offering to install non-condensing boilers. If they tell you any different it would be worth finding yourself a different plumber.

If your current boiler was installed around 1974 (and many still date from this time) it would only be 56 per cent efficient, 1985 efficiency jumps to 67 per cent and if you updated around

top tip

Lag that water tank . A basic insulating jacket costs as little as £14 but it will save you £29 a year, or 53 kg of carbon per year and pay for itself in just a few months. What are you waiting for?

fact

If everyone in the UK had a condensing boiler annual carbon dioxide emissions would be reduced by 17.5 million tonnes, saving £1.3 billion on energy bills every year.

2000, 74 per cent efficiency would be realistic. That 1970s boiler could be costing you almost 2 tonnes of extra carbon dioxide in heat alone each year.

If you replaced your boiler just before the 2005 regulation change it should be running at around three-quarters efficiency, in which case it wouldn't be financially or environmentally worth replacing – you'd be better off improving the overall efficiency of your heating system with thermostatic radiator valves (TRVs) and a good electronic thermostat (see Heating controls section for more details).

What should I buy?

All boilers are rated according to their energy efficiency by Seasonal Efficiency of Domestic Boilers in the UK (SEDBUK) and since 2005 only those rated A (90 per cent or above efficiency) or B (86-90 per cent efficiency) should be installed. Ask your plumber which boiler they recommend for your house and explain you want the most energy-efficient possible. You can double check they're offering you the best by visiting www.boilers.org.uk, which lists brands and models according to their energy efficiency. If you go into a plumbing showroom also look out for the Energy Saving Trust sticker of approval that guarantees efficiency of 90 per cent or above.

It may be worth considering a combination boiler (one that heats your radiators and produces hot water for baths and showers without the need for a separate water tank). Because this system doesn't need the hot water tank you should find installation costs are less, plus your hot water bill may go down as you're only using what you need rather than constantly filling a tank in the attic.

Additional savings can be made if you choose the size of boiler best suited to your home. There is no advantage in having a huge boiler: it will cost you more in fuel than the correct size, even working at the same level, so always check and double check with your plumber before choosing.

myth buster

'Condensing boilers are unreliable and difficult to install'

Fact:

Although condensing technology has been around for over 20 years –and in Germany 70 per cent of domestic heating systems use it successfully – the take-up in the UK was slow until mandated by the 2002 Building Regulations. Many industry commentators believe this has been due to an unwillingness to change old habits and suspicion of new technology in the heating and plumbing industry. While stories of difficult installations have surfaced, the truth of it is that often the only difference between a new and conventional system is the need for an additional drain for the condensate – and any competent installer should have no problem finding a suitable location for the drain in most circumstances.

how much you could save

Typical annual fuel costs (assumes 15% fuel price rises in January 2008)						
	Seasonal efficiency %	Flat	Bungalow	Terraced	Semi-detached	Detached
Old boiler (heavy weight)	55	£307	£392	£407	£457	£633
Old boiler (light weight)	65	£266	£337	£350	£391	£541
New boiler (non-condensing)	78	£227	£286	£297	£332	£455
Condensing boiler	88	£206	£258	£267	£298	£408

should I repair or replace my boiler?

'I've got a 9 year-old boiler that has never broken down, should I be replacing it with a new condensing boiler?'

Your current boiler will be running at around 70 -75 per cent efficiency – and this will only get worse as time goes on – compared to 90 per cent if you choose an A-rated design. If you can't afford to update your boiler, maximise its performance with a service and consider flushing out with a proprietory sludge remover that removes corrosive build-up that can reduce a boiler's performance by 15 per cent. But be warned, this will just be putting off the inevitable.

VAT'll do nicely:

Installation (including purchase) of most energy-saving products now attracts lower rates of VAT. If you have insulation, draught strips, heating controls, wood-fuelled boilers, solar hot water or PV, ground source heat pumps, wind turbines or micro CHP installed, you should only pay 5 per cent VAT, or none at all for a new build.

heating controls

What is it?

Most homes have two ways of controlling the central heating: a basic timer built into their boiler and a simple thermostat in a hallway or living room. While the latter will give a basic control of when the heat turns on and off and how warm the house is it's surprisingly inefficient. Significant savings can be made, by understanding how to use the basic controls well; upgrading the thermostat to a fully controllable electronic design that can manage your heating by measuring the temperature and assessing what the home needs to keep warm; and by adding thermostatic radiator valves (TRVs) to each of the radiators, so that each room can be controlled according to how often you're in them.

How much will it cost?

Understanding your heating controls costs nothing – in fact it should save you pounds. TRVs cost from £6.50 each (www.inspiredheating.co.uk) and are easily available from plumbing merchants and hardware stores. They can be fitted by a competent DIY-er (they simply replace the simple valve on your radiators). A new fully controllable heat programmer will cost at least £150-£200 but may require an experienced plumber to fit it.

How much cash will I save?

Expect to shave £60-£70 per year off your heating bills with upgraded controls. With TRVs you can make instant savings on fuel because in rooms you don't use that often (the spare room is an ideal example) you can turn the temperature down or off to limit the fuel you use. It is not a good idea, however, to have a TRV on the radiator in the same room as the main thermostat; this is because if it turns the radiator off at a lower temperature, it can mislead the main thermostat into thinking that the house is cooler than it really is.

Some electronic programmers can work out how warm your home is and adjust accordingly so that instead of having the heating coming on at, say, 7 am each morning regardless, it will only come on when the temperature falls to a certain level.

top tip

Studies have shown that most of us don't fully understand our heating and hot water controls. Get to grips with the settings:

Set the heating to come on 45 minutes before you want the house to be warm (like pre-heating an oven), and to go off some time before you go out or to bed.

Turn down radiators in bedrooms and in rooms that get a lot of uncontrolled heat – for example from the sun or from cooking.

When entering a cold house many of us turn the room thermostat up – there's no point: the boiler can't put out heat quicker just because the thermostat is set higher.

On warmer days it will come on later because it needs less time to heat the home. Even a basic electronic programmer will save you money because you can set a diary of heating rather than running a boiler for long periods.

How much carbon will I save?

The Energy Saving Trust suggests a combined saving of 280 kg per year of carbon dioxide while DEFRA has worked out that's a saving of 3.39 tonnes over the lifetime of the programmer.

What should I buy?

TRVs are readily available from DIY and plumbing merchants and you'll be able to choose a number of styles at different prices, but they all do pretty much the same job.

Intelligent heating controllers such as the Dataterm (£223, www.warmworld.co.uk) cost more than standard timers but can learn about the house and use sensors to remember how long a house takes to heat up in different weather conditions; you can also set them to keep the house at specific temperatures throughout the day.

top tip

Reflective radiator panels are cheap to buy, easy to install and reflect back heat that would otherwise escape through the wall. They can be bought from DIY stores or you can make your own by wrapping tinfoil around cardboard

biomass heating

What is it?

In a nutshell, biomass heating uses organic matter (wood is the main source) as fuel to heat your home. There are currently two effective ways to do this: **(1)** installing or using a space heater such as a wood-burning stove or a fireplace to heat the room it's in; **(2)** installing a biomass boiler that uses wood pellets, logs or chips for fuel. Using wood for heat (it's not a new idea) is one of the very few carbon-neutral ways of heating homes because trees only release as much carbon dioxide as they take up. As long as sustainable supplies are used, the total amount of carbon dioxide used (not including manufacturing and transport) will be as close to zero as possible.

There are many domestic log, wood chip and wood pellet burning central heating boilers available but some, like the log

top tip

For accredited biomass boiler installers visit www.lowcarbonbuildings.org.uk and for a list of suitable boilers try www.clear-skies.org.

boiler, may not be suitable for everyone as the wood has to be added by hand – not very practical. Wood pellet and chip boilers use automatic feeders that load the furnace with fuel as it needs it. These boilers can be extremely efficient and highly automated – some even clean themselves.

How much will it cost?

A 15 kW (the average size needed for a three-bedroomed semi) wood pellet or chip boiler costs between £9,000 and £19,000, a log burning boiler between £7,000 and £11,000 so this isn't a cheap option; but it is exceptionally green and removes a home's need for gas, greatly reducing its carbon footprint. Remember, you're responsible for the supply of fuel to the boiler. So instead of paying a direct debit to your gas company you'll pay per tonne or bag of fuel to a supplier. On average logs cost £65-£85 per tonne and wood pellets or chips £246 per tonne.

A cheaper alternative would be to open up an existing fireplace and install a wood-burning stove. You can pick these up second-hand from as little as £150. Although these would only heat one room you'll still be removing the need to use the radiators, and therefore the boiler to keep it warm. Open fires, while pretty, are not very efficient – it makes more sense to use a stove to get the most energy possible out of your wood.

How much cash will I save?

If you have good access to a wood supply and your boiler can provide you with all your heating and hot water, you could save £350 a year in bills. According to XCO2 (www.xco2.com) a house burning wood pellets would spend £373 per year compared to £550 for delivered gas – especially significant if you are not connected to a mains supply.

How much carbon will I save?

Because you're effectively removing the need for fossil fuels you could save 4 to 7 tonnes of carbon dioxide per household per year.

Is my home suitable?

Domestic wood-powered boilers are not yet suitable for all homes so check the following first:

Fuel: Do you have room for a large boiler and storage for wood chips, pellets or logs? (A separate room is often needed

fact

The average UK gas bill in 2007 was a whopping £579 per year. A highly-insulated efficient building such as the Bow Yard development in Somerset (see p.298) can cost just a few pounds to heat, which just goes to show how important insulation is in the fight against bulging bills and clod-hopping carbon footprints.

just for the boiler and fuel). Some suppliers will deliver your fuel on a truck, so you may also have to have suitable access for large vehicles. And check the availability of biomass fuel in your area – the further the fuel has to travel the less carbon you'll save.

Smoke-free zone: Under the Clean Air Act many areas are free from chimney smoke although some woodburners are approved for use in smoke-free zones.

Chimney: If you live in a listed building you may need planning permission to install the flue for the boiler; and remember, any work will need to comply with building regulations.

Location: It will be far easier to source wood in rural rather than city locations, and living in an urban area will almost certainly add to the cost of your fuel supply. If you are off the gas mains you will save more money and carbon by using wood.

What should I buy?

There are dozens of different boilers, many capable of 95 per cent efficiency (as good if not better than a gas-fired condensing design). It's worth talking to the experts to find out the correct size and fuel you'll need for your home. A small 2 kW boiler may well be enough heat for a very well-insulated home, so it's worth incorporating other carbon-saving features at the same time.

Where can I buy biomass?

Unless you're lucky enough to own a forest you'll need to find a fuel supplier that can deliver sacks or loose woodchips or pellets by the tonne. In the past the majority of pellets came from overseas, greatly increasing the carbon miles. This is slowly changing, however. For example, the companies Energy Crops and Biojoule are combining efforts to build a new pellet mill in Nottinghamshire that will produce 10,000 tonnes of pellets a year.

Remember the simple things. Turning your thermostat down by 1°C can cut up to 10 per cent off heating bills. Go on, you'll never notice the difference.

bags and bags of biomass

Another great advantage of wood chip and biomass heating systems is that they don't always rely on virgin wood resources. In fact there is a great deal of waste wood (often destined for landfill) that can be easily processed for heating. Barnsley Metropolitan Borough Council made the most of the council's waste wood by installing biomass boilers in several blocks of council housing. This has enabled a small wood-chip supply business to start up that re-processes waste without the need for additional trees.

case study
'a wood burner works for us'

'I read an article a while ago [about] the huge amount of wood that was sent to landfill every year. I think it was something like 90,000 tonnes. I did a bit of research and realised that in order to heat our house and hot water via a log-burning stove would use about 3 to 5 tonnes of wood a year. We'd just moved into a 1950s ex-council house that needed a complete overhaul. It already had a gas back-boiler and so was perfect for conversion to a log burner. After several weeks on Ebay I found a log burner specifically designed for council houses and a quick bid and a trip to Nottingham later we were suddenly the proud owners of a log burner. Once I started looking around and asking I found that there's a huge amount of timber being thrown away every day. This relatively simple act is our biggest commitment to date to cut our carbon emissions and I think is easily achievable by many families in the UK.'

David Denbury, Friends of the Earth supporter

HOUSEHOLD APPLIANCES

how much carbon do my white goods produce per year?

top tip

Keep fridge and freezer doors closed. Each minute a fridge door is open it can take three energy-hungry minutes for it to cool down again

Domestic appliances, such as fridges and cookers, are responsible for nearly half (47 per cent) of the total electricity consumption in our homes and on average around 1 tonne of carbon dioxide emissions a year. In other words, we spend nearly half our electricity bill on running appliances, so choosing the most energy-efficient makes perfect sense. But with Curry's selling 50 different fridge freezers and John Lewis 55 different washing machines, knowing what to choose isn't always easy.

This section explains what you need to know, the questions you need to ask and how to use your appliances (or not) to keep their carbon contributions to a minimum.

Typical annual carbon dioxide emissions for kitchen appliances

Appliance	Most efficient (kg CO_2)	Typical new (kg CO_2)	Typical old (kg CO_2)
Fridge freezer	72	188	301
Chest freezer	41	140	233
Upright freezer	51	153	224
Refrigerator (with icebox)	39	114	156
Washing machine	86	96	128
Dishwasher – full size	113	131	199

quids in

'I received a hefty bill from my electricity company for almost £300 based on an estimate of what we'd been using. I dutifully phoned them with a proper meter reading, hoping for an overestimate. Holding my breath for a second, I asked the fella on the other end how much we owed them. I could hear a sharp intake of breath and he asked me to repeat the meter reading. I repeated the numbers. He coughed and said, 'Actually, Miss, *we owe you money*.'

So that's proof that switching things off at the wall and not leaving lights on etc can wipe £300 off your electricity bill in an instant and help save the planet at the same time. Quids in.'

Sophie Powell,
Friends of the Earth supporter

typical electronic appliance CO_2 emissions

Consumer electronics	Annual CO_2 emissions (kg)
14" CRT in bedroom	33
42" plasma TV	439
Broadband router	55
Compact Hi-Fi	43
Digital radio	4
DVD home theatre	40
Freeview set top box	30
Laptop	45
PC & monitor	150
PC speakers	15
Printer	50
PS3	115
Scanner	25
Sky +	84
VCR	56
Xbox	71

'I can't afford to replace my fridge'

If your appliances are getting old but still work well, consider buying a Savaplug (www.savawatt.com). This plug-in gadget reduces the flow of electricity to your fridge to match the actual amount it needs, making it up to 20 per cent more efficient. They cost £24.99, but make sure you check your model is compatible. Many modern fridges now incorporate such a feature. Always go for an A or A+ rated fridge.

energy efficiency recommended logo

This symbol is your guarantee that the appliances you're buying will save you energy, cost less to run and reduce the amount of carbon your home produces. Choose A, A+ or A++ appliances to ensure you make the biggest savings. Be warned, though: each appliance is graded in its own category, so you can still buy a huge energy-sucking American style fridge freezer with an A rating. So look beyond the energy rating and see exactly how much electricity an appliance consumes.

Energy Washing machine

Manufacturer
Model

More efficient
A
B
C
D
E
F
G
Less efficient

Energy consumption kWh/cycle (based on standard test results for 60°C cotton cycle) Actual energy consumption will depend on how the appliance is used	0.95
Washing performance A: higher G: lower	A B C D E F G
Spin drying performance A: higher G: lower	A B C D E F G
Spin speed (rpm)	1400
Capacity (cotton) kg	5.0
Water consumption /	55
Noise (dB(A) re 1 pW) Washing	5.2
Spinning	7.0

Further information is continued in product brochures

Energy efficiency recommended logo

EU eco-label

This voluntary labelling scheme is also a good indicator that the product will save you energy. To earn the label, products must meet a range of criteria beyond energy consumption alone. These can include:

- commitment to recycle at the end of its useful life
- A-rated energy efficiency, or better
- no flame retardants harmful to the environment

EU eco-label

kitchen appliances

Here's a carbon-friendly run down of the main appliances used in the home.

dishwashers

What to look for:

Dishwashers are rated according to energy and drying. Choose an A rating for both to get the best environmental performance. Waterwise (a member of the government's Water Saving Group) lists dishwashers according to water consumption (www.waterwise.org.uk). At the time of writing the Miele

G1530SC full size is the best, using just 10 litres per cycle; however Bosch claims its Bosch Logixx SGS65L22 uses just 9 litres during one of its cycles.

Energy-saving tips:

Dishwashing efficiency hasn't changed enormously over the past few years so if your machine still works you can improve its efficiency by ensuring it's full, reducing the number of times you use it, and using the coolest setting possible.

top tip

Pack the freezer. It takes less energy for a full freezer to stay cool than it does an empty one. Don't have enough food to fill it? Use plastic bottles filled with water.

dishwashers versus washing up by hand

Jacob Tompkins of Waterwise says, 'Washing up uses about 10 per cent of the water consumed daily and using a modern dishwasher could reduce this to less than 2 per cent – if every household in the UK did this it would save more than a quarter of a billion litres a day.'

According to research verified by the University of Bonn the average household will do two or three hand washes a day and uses 60 litres of water, while a new dishwasher uses on average 12 litres per wash. Appliance manufacturer Bosch, keen to promote its eco-credentials, claims hand washing would account for ten times as much water as a dishwasher.

What this doesn't take into account, however, is the total energy used to manufacture and run the dishwasher, the transport costs and raw materials used or hand washing habits.

fridges and freezers

What to look for:

Look for A+ and A++ energy labels for super-efficient designs that use 23-46 per cent less energy than a standard A-rated model. They are more expensive and choice is still limited compared to A-rated, but things are improving. AEG/Zanussi, Miele and Liebherr all have a good selection. The Energy Saving Trust (EST) calculates savings in running costs of £45 per year and 185 kg carbon dioxide for A+ fridge freezers compared to a non A-rated appliance.

Always choose hydrofluorocarbon (HFC)-free models. HFC coolant has a global warming potential of around 3,200 times that of carbon dioxide. Look instead for the R600a hydrocarbon coolant label as this has a lower global warming potential, is non-toxic and is also more efficient than HFCs.

Energy-saving tips:

Frost-free is a popular and convenient extra on some models, but be warned, not having to defrost comes at a price – up to 45 per cent more energy is needed for these models.

washing machines

What to look for:

Washing machines are rated on energy, spin and wash so choose AAA designs. A+AA is also becoming more available, meaning you get even greater energy savings. It's particularly worth choosing the highest spin speed you can (1400 rpm +) because an A-rated spin cycle will leave 3 kg of dry clothes with no more than 1.3 kg of water compared to a C-rated design that would take considerably longer to dry on the line because it will still hold up to 1.89 kg of water. It may take a little more energy to spin your clothes but this would be considerably less damaging than having to leave the heating on for an extra few hours to get your clothes dry.

Energy-saving tips:

Wash everything at 30-40°C – you won't notice the difference in cleanliness because most of the time clothes only need a quick refresh rather than a deep clean. Unless your washing machine has fuzzy logic sensors that adjust the amount of water used for each wash, your machine will use the same amount of water and energy if it's cleaning 6 or 16 shirts.

tumble dryers

What to look for: If you really can't live without one, spend a little more and choose one of the few A-rated designs such as the AEG T59800. This uses condensing technology and a heat pump to save half the energy that would otherwise be wasted – 2.1 kWh compared to 3 kWh or 4 kWh for C-rated models.

But remember: the cheapest, most energy-efficient clothes dryer is a clothes line. Think very carefully before installing a dryer in your home because few items use so much energy to do something that can be done naturally.

top tip

Ninety per cent of the energy that washing machines use goes to heat the water, so switch to a cooler wash: today's washing powders work just as well on low temperatures.

Energy-saving tips:
If you have to use a tumble dryer always fill to its optimum level, and avoid overfilling as this will take longer to dry, using more energy.

ovens

What to look for:
Opt for an A-rated electric fan oven with at least a double-glazed door (triple are available on some) and always check the kWh rating to pick the lowest rate possible within the A band – between 0.6 kWh and 1.0 kWh depending on the capacity of the oven. If you are buying a double oven check that both cavities are A-rated. Often the second smaller oven is B- or even C-rated.

Energy-saving tip:
Multi-task the oven to make the most of the heat. If you're baking a cake, think about what else you could cook at the same time. If you can squeeze in a crumble as well you'll save yourself time and energy and have a spare pudding for the freezer.

What about microwaves?
These use 50-70 per cent less electricity than a conventional oven, making them extremely efficient. But regardless of what the box says they are not yet good enough for all cooking tasks. This makes them a really useful tool for cooking fish and vegetables, heating, melting, pre-cooking and re-heating when turning the main oven on would be wasteful. If you have one you probably can't live without it, but if you don't you'll probably never need one.

top tip

Check the energy consumption on the EU Energy Label. This will show exactly how many kWh electricity an appliance will consume. Look for the lowest: for instance both the Hotpoint RLA34P (152kWh/year) and Bosch Logixx KTR18P20 (117kWh/year) fridges are A-rated but the Bosch uses 35 kWh/year less energy.

the SUV of the kitchen

Huge double-door American-style fridge designs use on average double the energy of a standard fridge freezer, and that's for an A-rated design. Over the course of its life this represents roughly the total electricity consumed by the average UK household in two and a half years. They might look fantastic but do you really need that much fridge space?

hobs

What to look for:

Hobs don't as yet require an energy-efficiency rating, which makes choosing one tricky, but you have a choice between three types: gas, ceramic and the latest induction hobs. In a comparison test in which 1 litre of water was boiled on an example of each, made by appliance manufacturer AEG Electrolux, ceramic hobs were found to consume 0.18 kWh, gas 0.26 kWh and induction just 0.13 kWh. Induction hobs use a magnetic field to heat ferrous metal pans (you may have to replace your old pans for stainless steel or cast iron designs) and because it is the pan and not the hob that heats up there is very little wasted energy. This may all sound a little complicated, but the upshot is you get the speed and controllability of a gas hob with the ease of use and cleaning of a ceramic/glass design. Induction hobs are more expensive, with prices starting at around £500, compared to gas and ceramic, which range from £200-£900.

Energy-saving tip:

Use the correct size pan for the ring to maximise its efficiency, always cook with a lid on, and use the least amount of water possible. Use a stacking steamer to cook more than one item at once on one hob. Invest in a pressure cooker - they require half the energy of a conventional pan.

top tip

Unless all your electricity is from a renewable source, gas is more efficient for cooking because it's a primary fuel rather than one that's produced in a big old inefficient power station miles away.

case study

keep-fit cookery

'Instead of electric gadgets I use a mouli-legumes to blend my soups and make sauces, I have a traditional hand whisk, a hand-pumped frother for milk and a hand-powered mincer. Think of all those extra calories burnt because I don't plug anything in.'

Penny Walker, London

using existing appliances

Being carbon neutral doesn't always mean having to shell out on new appliances. Many savings can be made by changing the ways we use the ones we already own. Here are some examples:

kettles

In Britain we drink 229 million cups of tea and coffee every day. In fact, about a quarter of all electricity used in domestic cooking is consumed by electric kettles. It's unthinkable that we could make do without. But rather than go out and buy an eco-friendly kettle the common-sense approach is to change the way you use your existing one. As the Department for Environment, Food and Rural Affairs (DEFRA) points out: 'If everyone boiled only the water they needed to make a cup of tea instead of filling the kettle every time, we could save enough electricity to run practically all the street lighting in the UK.'

That said, there are a couple of features you should look for when you next need to buy a kettle. Always choose a concealed element design – traditional exposed elements need to be fully submerged for them to work properly, which in some models means you can't boil a single cup of water. Avoid keep-warm functions and swanky neon lights (or worse, models with standby) – all are just unnecessary energy wasters.

the eco kettle

On the face of it the Eco Kettle (www.ecokettle.com) is a great idea. With the push of a button you can control the amount of water (one, two, three cups and so on) that goes into the boiling chamber. In a test it was calculated that if we all used the Eco-kettle it would save 697,000 tonnes of carbon dioxide annually compared to a standard design. Great stuff – but wouldn't the same savings be as easily achieved if we measured exactly how much water we needed (by filling the mug and pouring it into the kettle) with our existing kettle? Do it this way and you'll save around £40 (the cost of the Eco Kettle) – enough to add thermostatic radiator valves to four or five radiators around your home.

recycling appliances

Since 2 January 2007 the law has required electrical manufacturers to take responsibility for what happens to the products they sell at the end of their lives and now have to recover and recycle between 70-80 per cent of the products they make. The Waste Electrical and Electrical Equipment (WEEE) directive aims to reduce the 1 million tonnes of electronics and appliances that end up in landfill every year. The upshot of this is that you can now get your old appliances, TVs, videos, computers and the like recycled for free by the manufacturer. You can return the items to the shop you bought them from and they'll recycle it for you if you buy a replacement, take the appliance to a WEEE collection point (www.wasteconnect.co.uk for locations) or request the manufacturer picks it up from your home.

say no to standby

Our homes are overflowing with electronic equipment, almost all of which will have a standby or sleep mode. Devices left in sleep mode, rather than being turned off, waste about two power stations' worth of electricity each year and produce the same amount of carbon dioxide as 1.4 million long-haul flights or the climate impact of about half a million flights.

Estimated annual carbon dioxide emissions in the UK from devices left on standby

Stereos:	1,600,000 tonnes
Videos:	960,000 tonnes
TVs:	480,000 tonnes
Consoles:	390,000 tonnes
DVD players:	100,000 tonnes
Set-top boxes:	60,000 tonnes

When you switch a modern electrical appliance off, it rarely actually shuts down: this only happens if you unplug it or turn the socket off. Some well-engineered devices draw just milliwatts of current in standby, whereas others continue to suck up 30 watts from the mains, or 113 kg carbon dioxide per year without it doing anything.

castaway:

The WEEE man takes his name from the Waste Electrical and Electrical Equipment directive – a European regulation designed to tackle the growing mountains of electrical stuff thrown away. He is built from the amount of electrical and electronic waste the average British person creates in their lifetime – he weighs 3.3 tonnes. It's not just the waste involved, the chucking of landfill and the lack of recycling that need to be taken into account with each appliance we buy (and then dispose of), but also the enormous amount of energy and emissions that went into its production. Of each kilogram of carbon dioxide emissions produced, around a third comes directly from fuel used during manufacturing; so before your appliance even gets to the shop, let alone into your kitchen, it's already responsible for a great deal of carbon dioxide emissions. That's why it's so important to only buy new appliances when there's no other alternative, and to make sure that old goods are properly recycled. Using recycled steel consumes 75 per cent less energy, for example, than extracting the raw material.

making life easier?

Smoothie-makers, steamers, bread makers, ice cream machines, toasted sandwich makers, lean mean grilling machines, popcorn makers...all make our lives easier don't they? They might look good on the worktop but before you buy stop and think. Can I already make that snack without it? Will I really use it?

using energy-saving gadgets

Standby plug:
This clever little multi-plug adaptor costs around £16 and takes the hassle out of switching all your electrical equipment off at the wall. Plug your computer into the master socket, and anything connected to the other sockets will be turned off when you shut down or switch off, and turned on again when you fire up your computer. It also works with hi-fi, TVs and energy-hungry set-top boxes.

A cheaper, if a little more inconvenient, option would be to spend a few seconds to turn everything off at the wall. Use a multi-socket adaptor and you could shave seconds off by only needing to flick one switch.

e-cube:
This helps regulate the amount of energy your fridge uses. Fitted over the fridge thermostat, it mimics the temperature of the food rather than the air in the fridge – which warms up every time you open the door. That way the fridge keeps food at the right temperature, but uses less energy to do so. Tests in hotels and supermarkets – where fridge doors are opened frequently – have shown energy savings of up to 30 per cent. With 87 million fridges in the UK, such savings could add up.

Bye Bye Standby:
A device that allows you turn off standby buttons by remote control, helping to make sure you turn everything off. Its makers estimate it can save 10 per cent of electricity bills, reducing emissions as well. Intelliplugs™ do a similar job, turning off stereos, computers and other plugged in accessories at the flick of a switch.

electronic appliances

Here's a carbon-friendly run down of some of our most popular electronic appliances.

computers

The average PC takes 1.8 tonnes of chemicals, fossil fuel and water to manufacture and causes the emission of 100 kg carbon dioxide each year, but most of us simply couldn't live without one. Unfortunately, as the amount we use them increases, as does the number of programmes we use, the

WHITE HEAT:
Some computers consume more electricity than others – as much as forty times more. If you're choosing a gadget check how power hungry it will be.

home computer simply gets bigger, and consumes more power to cope with the demand. This is especially true of desktop PCs that are generally plugged in all day. (Unless you flick the switch on the wall or take the plug out of the wall a PC that is apparently turned off will still be using power). And if you leave your computer monitor on all night you'll waste enough energy to microwave six dinners. A PC left running 24 hours per day would use £59 worth of electricity over a 12-month period and create 716 kg of carbon dioxide emissions a year. Laptops, on the other hand, need to be as energy-efficient as possible so that they can go anywhere. The processors are designed to run on less power, the screens use as little energy as possible, all of which adds up to significant energy savings compared to a desktop PC.

Energy-saving tips:

Unplug your PC when not in use, and don't forget the scanner, monitor, printer, broadband box and audio speakers all need turning off too. Use your computer's energy saving mode: you should be able to turn the brightness of the screen down, or set it to turn off if you haven't used it for 5-10 minutes. For instructions on how to do this on PCs or Macs visit www.nrel.gov/sustainable_nrel/energy_saving.html.

TVs

The next time you go to buy a TV the chances are that, unless you're looking for second-hand, you'll be choosing between either an LCD or a plasma screen, rather than the traditional cathode ray tube (CRT) box. It's easy to see the appeal of the new-style televisions: they're sleek, sexy and space-saving. But plasmas also consume more energy than the old CRT models. A CRT can use as little as 90 W of electricity compared to up to 390 W for a plasma screen or 270 W for LCD. But for the equivalent size LCD is better than CRT. With screen sizes getting bigger, however (42-inch screens are fast becoming the most desired), so the energy use increases. Currently 92 per cent of TVs are CRT, but this figure is expected to be 52 per cent by 2010. And by 2020 they're expected to be a thing of the past.

What to look for:
The most sensible energy-saving sets are known as IDTV (integrated digital television). These have a digital set-top box built in, meaning you get two appliances but only one plug, and unlike many set-top boxes they can be switched off completely without losing any settings that have been made. Choosing an IDTV might only save you 25 kg of carbon dioxide a year, but the combined savings across the nation's 60 million TV sets would be close to 1.5 million tonnes of carbon dioxide. DEFRA estimates that there could be 80 million set-top boxes in the UK by 2010.

set-top boxes

At the time of writing there are 2.6 televisions per household in the UK. With the switch to a digital signal due to be under way in 2008 many of these TVs will need upgrading in order to receive the digital signal. The cheapest and easiest way to do this is through buying a set-top digital receiver. It's estimated that by 2010 there will be 80 million of these energy-hungry devices and many of them need to be kept on standby for them to work properly. Sky – purveyor of the always-on Sky+ boxes – has begun to address this problem: its boxes now automatically shut down into a deep standby after 11pm each night.
What to look for:
Check both energy consumption and standby rates and always choose the lowest. But be sure that the box can, at the very least, be switched into deep standby, or ideally turned off all together. Look out for the Energy Saving Recommended logo.

www.sust-it.net

This useful site lists thousands of appliances and electrical goods according to how much energy they use. It will also show how much an appliance costs to run per day, per year and while in standby. Some surprises include:

- The Apple Mac mini might be small but it still uses a power-guzzling 110 W to work and costs £39 a year to run compared to the Very PC 478 that uses just 59 W and costs £21 per year.

- The Bosch WLX24162 Washing Machine uses just 0.76 kWh/cycle or £32.38 a year compared to the Hoover VHD816 that uses 1.52 kWh/cycle and will cost you £70.55

- Sony PlayStation 3 (PS3) Video Games Console (US Version) uses a whopping 380 W and costs £43.77 a year to run compared to the Nintendo Wii Video Games Console that needs just 17 W to work and costs just £1.96 a year.

LIGHTING

Nearly a third (30 per cent) of our electricity bill pays for our lighting. That's the equivalent of a staggering 19 per cent of total global energy. In other words, our pendants and chandeliers, up and down lighters, dimmers, table, side and floor lamps consume an incredible amount of energy. But they don't have to. The International Energy Agency has worked out that if the world switched to low-energy light bulbs it would save 16 billion tons of carbon dioxide, not to mention saving US$2.6 trillion (£1,300 billion) in electricity bills.

It's taken most of us a while to get used to energy-saving bulbs. Early complaints were 'The light's poor, they flicker and they're too expensive.' But the technology has improved enormously and there's now a great selection available at affordable prices. Philips has been selling standard Compact Fluorescent Lamps (CFLs) for just 49 pence each and Tesco said it wanted to sell 10 million energy-saving light bulbs between April 2007 and April 2008.

top tip

Turn lights off. Lighting an empty office overnight can waste the energy required to heat water for 1,000 cups of coffee.

low-energy light bulbs

What are they?

Incandescent light bulbs (the classic filament design) have changed very little since their invention in 1879. Around 95 per cent of the energy they use is wasted as heat with only 5 per cent going towards lighting our homes. The average incandescent bulb lasts just five months to a year. Modern low-energy bulbs (CFLs), on the other hand, require only 20 per cent of the energy incandescents need to produce the same amount of light and can last up to 12 times longer.

How much will it cost?

With special promotions, Buy One-Get One Free offers and multi-pack deals prices can be as little as £0.49 per bulb (Philips ranges from John Lewis). And although the average price for a single CFL is around £3.50, this will fall as demand increases. The average home has 28 lights, so replacing all the bulbs could cost as little as £14 but as much as £98. Top tip: shop around.

How much cash will I save?

Research by Greenpeace UK suggests an average 15 per cent saving on electricity costs, or around £50 per year simply by changing to CFL bulbs. The Energy Saving Trust compared a standard 100 W bulb and its equivalent 18 W CFL bulb and over their lifetime the CFL cost just £18 (10,000 hours) compared to £73 (just 1,000 hours life). That's a £55 saving per bulb.

How much carbon will I save?

As a country, switching from incandescent to CFL bulbs for domestic use could see savings of up to 5 million tonnes of carbon dioxide – the equivalent of emissions from a medium-sized coal-fired power station. A CFL produces 70 per cent less carbon dioxide than an old-fashioned bulb and will save around 38 kg of carbon dioxide over its lifetime.

What should I buy?

Like appliances light bulbs are graded A-G (A being the best) according to their efficiency, so it makes sense to buy A to maximise the savings you'll make. As for choice, Homebase sells dozens of different types of low-energy light bulbs. You can also order them from catalogues such as The Natural Collection.

top tip

Lap up natural light. Light-coloured walls, ceilings and floors, as well as mirrors reflect daylight, reducing the need for artificial lighting.

let there be light

Before you reach for the light switch...

Open the curtains and blinds – you can't beat natural light, and it's free.

Clean the windows – you'd be amazed at how much more light comes through a sparkling pane.

Dust the lampshades – lampshades and bulbs covered in dust won't light half as well. A quick dust will be an illuminating experience.

light without electricity

All homes, unless you're lucky, have a dark area that needs constant lighting. Maybe it's a windowless bathroom or box room or a long corridor that lacks natural light. Well, instead of installing bright halogens to light the way consider light tubes (Solatube or Sunpipes) that draw natural light from a small transparent dome on the roof down a highly reflective pipe and into your home. The results can be stunning, flooding a dark corner with light. With prices starting from £180 they're reasonable value too. It is worth checking what the manufacturer says about any effects this might have on your insulation.

energy-saving myth busters

They can't be used with dimmer switches

Fact

A standard CFL can't cope with the change in voltage caused by turning a dimmer and will flicker, but improvements are coming. DigiFlux has developed the first dimmer-proof CFL. At £11.99 each, they're more expensive, but as the technology improves, prices will tumble. Dimmable designs are also available that can be controlled from a standard light switch. Prices start from around £10.

Turning CFLs on and off consumes more energy than leaving them on all day

Fact

This is an age-old myth referring to old-fashioned strip lighting that needed lots of power to warm up the tube. Modern low-energy bulbs use almost no extra electricity to warm up – so switch them off just as you would an old-fashioned bulb.

switch to green electricity

Switching to a green electricity tariff – one that supplies your house with power from clean renewable sources such as wind or hydro – is a quick, affordable and virtually painless way to cut your carbon emissions by as much as 2 tonnes. Or is it?

Only a tiny percentage of UK electricity at present is produced from renewables. If we're to cut emissions significantly there needs to be a big shift to renewable – and that's going to require big investment. Switching to a green electricity supplier will send out a message that you want to do your bit; and it adds to demand in the market for green energy. But it's only a win for the planet if it really pushes supply companies to add new green generating capacity.

Via the Renewables Obligation (see Chapter 6) the government has ordered suppliers to get a proportion of their energy is from clean, green sources. Companies generating more than the minimum can sell any surplus in the form of certificates to rivals that haven't met their targets. The danger is that you buy green electricity – but someone else just gets dirtier electricity.

Some companies take certificates out of the market, making it harder for other companies to meet their own green targets the polluting way, and helping force them into investing in more renewables.

What happens when you switch to a green tariff? Nothing at all – you don't need a new meter or any special equipment. It happens just like that.

worth considering:

Good Energy – 100 per cent renewable electricity. Good Energy also encourages customers to install microgeneration, by paying a fee for every kWh of renewable electricity generated, whether it's used by the customer or fed back into the grid.

Green Energy 100 – For every unit of electricity the consumer buys, Green Energy 100 buys one unit of green electricity. Green Energy 100 also invests 50 per cent of profits into green electricity generation.

Ecotricity – Invests heavily in wind farms, spending more per customer than any other company

Guidelines on green tariffs are changing, so check out www.whichgreen.org for up-to-date information. Change suppliers at www.uswitch.com

FLASH: don't bin your bulbs

Energy-saving bulbs contain a small amount of mercury, which is highly toxic. Although not all local councils offer CFL recycling, don't be tempted to just bin them. Thankfully there are a number of shops, including Ikea, that will recycle them, so collect up any that you have, ask friends or neighbours to do the same and take them the next time you go. Companies such as Recyclite (www.recyclite.co.uk) will also collect and recycle CFL bulbs.

In March 2007 the government announced it would ban incandescent bulbs by 2011; and the introduction of the WEEE directive in January 2007 should mean improved recycling facilities to cope with the increase in CFLs. Some local councils will accept bulbs for recycling, so it's worth contacting them first to see where your nearest recycling centre is.

LED-ing the way?

Light-emitting diode (LED) bulbs won't burn out, they use a tiny amount of energy (ten times less than incandescent bulbs), and will last up to 30,000 hours. At present domestic LED lighting is used mainly for decoration rather than as the main source of light for a room, but new (and rapidly improving) technology means they are now starting to become available for everyday use. Be warned, however: prices are high and reports suggest the light isn't yet as good as it needs to be. But that's what they said about CFL bulbs a few years ago, so watch this space.

what about low-voltage lights?

Don't get confused between low voltage (typically halogen bulbs) and low energy CFLs. While these bright bulbs (mostly found in the kitchen) are 10-20 per cent more efficient than a conventional light bulb, they still use a great deal more than a CFL, especially as they are usually displayed in rows of three or

how do your bulbs perform?

(Based on a 60 W bulb, 11 W CFL, 12 V LED GU10 fitting, 12 V halogen GU10 fitting)

	CFL	**Incandescent (traditional)**	**LED**	**Halogen**
Cost	£3.50	£0.50	£9	£1
Running cost	£2.41 per year	£13.14 per year	£0.4 per bulb per year	£14 per 75 watts per year
Lifespan	6-8000+hours (five months)	750-1,000 hours	30,000 + hours	2,000-3,000 hours

four, compounding the problem. You can, however, get compact CFL bulbs that will work in the standard halogen GU10 fitting. Bulb manufacturer Megaman produce a range which last seven and a half times longer than halogen bulbs with up to 80 per cent energy savings. www.megamanuk.com, stockists 0845 408 4625. If you can't get the right CFL, it's worth knowing that manufacturer Osram makes energy-saving halogen bulbs that use around 30 per cent less, but with the same output as their energy-hungry equivalents. See www.osram.com.

case study

'I've saved myself £11,000'

Rod Webb owns the Green Man pub in Milwich, Staffordshire, and after repeatedly replacing conventional light bulbs he changed all 28 in the public bar to low-energy varieties. The next electricity bill he received allowed him to halve the direct debit to the supply company. Over the 16 years that he has been using low-energy bulbs Rod estimates that he has used 100,000 kWh less electricity than he would have done with ordinary bulbs – that's a saving of £11,000. Had he been using ordinary bulbs, he would have had to change over 1,300 light bulbs.

Rod says: 'The only problem I have is the lampshades. With ordinary bulbs I used to clean them each time I changed the bulb, now the bulbs last so long I have to remember to clean them.'

Rod Webb is a member of Friends of the Earth, Stafford

KITCHEN ETIQUETTE

Food production and agriculture are estimated to be responsible for around 8 per cent of the UK's total carbon dioxide emissions. But globally the food system – which includes farming, deforestation, producing cattle feeds and fertilisers, and food processing – accounts for about a third of all greenhouse gas emissions. In fact almost all our non-carbon dioxide emissions come from agriculture.

But it's pretty easy to reduce the carbon in your diet, and often as simple as just asking 'Where does it come from?' So what questions should you be asking, and what answers should you be looking for?

start a low-carbon diet

Follow these simple steps to ensure your groceries are as carbon-neutral as possible.

Eat less meat. According to the United Nations Food and Agriculture Organisation, the global livestock sector generates more greenhouse gas emissions (including methane) than all forms of transport, so cutting down on or giving up meat and dairy will slash your footprint. When you do buy meat try to go for organic, locally produced or free-range from your local butcher. And remember that pork has lower global warming impact than lamb or beef.
Eat lower down the food chain. Cereals and veg tend to have less impact on the climate than meat and dairy.
Always read the label. The nearer to home the food was grown or manufactured the less carbon it will have taken to get it to your shopping basket. If there's no label, ask the shop keeper.
Eat seasonally. Check out www.eattheseasons.co.uk to see what's available each month.
Grow your own. Then you're talking food metres not miles. There may be local community growing schemes if you're lacking a garden or window ledge. Garden Organic can help (www.gardenorganic.org.uk).

Go to your local farmers' market and support regional producers. Try www.farmersmarkets.net; and make the most of your local farm shops at www.farmshopping.com.

Buy Fairtrade. For non-European foods, choose certified Fairtrade products – they help communities develop out of poverty.

Eat organic or extensively reared/free range food, and avoid artificially fertilised produce (see www.soilassociation.org).

Avoid processed and packaged foods. Enormous factories chop, blend and package your oven chips. And more is processed overseas than ever before. What's wrong with a potato, a knife and an oven tray?

Accept ugly. Mountains of food gets thrown away because it doesn't look right. Imperfect-looking tomatoes can still taste perfect.

Compost food waste. A compost heap in your garden will produce fantastic feed for plants and veg. If you don't have a garden your council is likely to supply you with a compost bin that they'll collect. If they don't, ask why and demand it.
Drink tap water. It's filtered just as much as bottled brands and doesn't need to be shipped in from the foothills of a glacier. Keep a chilled bottle in the fridge, and if you can't live without carbonated water try a Soda Stream and fizz up your own.

slash your food miles

You don't have to go without those exotic favourites. Many delicacies made famous in far-flung places are now grown or made virtually on your doorstep.
Chillies from Cumbria: Seventy varieties of the spicy pepper with more than 100,000 chillies grown yearly and turned into delicious sauces. www.treescantdance.co.uk
Chinese pak choi from East Anglia: The classic stir-fry vegetable grows perfectly in Cambridgeshire.
Champers from Sussex: There may only be 78 hectares of vines but they produce a fizz that rivals the best Champagne. www.nyetimber-vineyard.com
Parma ham from Dorset: Italian finest cured slivers or dried salted pork with a British twist from Den Hay farms. www.denhay.co.uk
Welsh soy: Handmade Shoda sauce from Abertillery near Gwent. www.shodasauceseu.com
Feta from Yorkshire: Fine Fettle Yorkshire Cheese has the same crumbly texture as its famous Greek cousin. www.shepherdspurse.co.uk
Hackney salads: Pick up delicious exotic salads from the heart of east London as part of a local weekly veg box delivery scheme. www.growingcommunities.org

BUYING AN ECO-FRIENDLY HOME

A house is so much more than bricks and mortar. It's the place where lives are lived, families started and relationships forged. A place to feel safe and where you are in control. It's no wonder then that buying one can be an emotional rollercoaster – even before the estate agents get involved.

Large kitchen...check. Three bedrooms...check. Close to good schools...check. New bathroom suite...check. But what about the impact the new place is going to have on the environment? And what about the insulation, boiler efficiency, heating controls or orientation?

How a house looks is obviously important – after all, you want to enjoy living there – but the way the building works will determine how comfortable you are in it. A new bathroom suite might be a pleasure to use and really catch the eye, but with poor insulation and rattling old windows the room will be so cold in the winter you may never get to spend more than a couple of minutes in there.

So what should you look for when making your move on to or up the property ladder?

buying an old house

As part of the government's drive to make house-buying quicker and simpler every home on sale since 2007 will need a Home Information Pack (HIP) containing an Energy Performance Certificate (EPC). This shows at a glance how efficient a house is and what its carbon emissions are. It will also list energy-saving recommendations and show how much more efficient the house could be. Just like a washing machine our homes can have a rating from A to G, with A being the most efficient.

So when you're faced with a dozen houses to view, let the EPC make some decisions for you and reject the least efficient houses on the list – or at least bear in mind that you'll need to spend some money to get them up to standard. This will not only make your future home less of a carbon guzzler, but will encourage the sellers to do something about their property. If simple low effort, low-cost measures such as changing to efficient light bulbs can increase the appeal and actually add to the property value it makes economic sense. Faced with two neighbours selling identical homes, one with an E energy rating (average carbon emissions: 13 tonnes per year, heating costs: £1,173) and the other a C (average carbon emissions: 4.9 tonnes, heating costs: £457), which one would you buy?

what to look for – viewing houses with green-tinted spectacles

Insulation: What type of brickwork does the property have? Does it have cavity walls; if so, are they insulated? What's in the loft? Poke your head through the hatch and have a look. How thick is the insulation (you're ideally looking for 270 mm of mineral wool). How much double-glazing is there? What about draft excluders?
Heating: When was the boiler installed? Remember, after 2005 it has to be a condensing design; any older than ten years and it'll need replacing sooner rather than later. How is it controlled? Look out for a room thermostat (small box with a dial, often found in the hall) at the very least or an electronic timer and individual thermostatic radiator valves (larger-than-normal radiator controls).
Kitchen: Is there a gas or electric hob? If appliances are built in, are they A-rated or above?
Position: Which rooms get the most sunlight? Visit the property at different times of the day to see how much you'll benefit from free heating from the sun.
Location: Can you walk to the local shops, services and – just in case you haven't already factored this in – schools? How dependent on a car will you be? What about public transport? Could you cycle or walk to work?
Eco-features: Do the owners generate their own power with wind, solar or ground-source heat pumps, or maybe they recycle their waste water? If they have renewable technology ask how well it works and what savings they make on their bills. Check carefully that there are high levels of insulation and energy-saving measures in place already. Without them the wind turbine or solar panels might just be eco status symbols.

fact

According to a 2006 survey seven out of ten Brits believe energy-saving features are worth paying extra for – and who can blame them? Think of the energy-saving home as being a bit like the low-energy light bulb. It might cost a little more up front but every day you use it, it saves you money and energy and reduces your carbon footprint. If you were guaranteed half-price electricity forever for just £2,000 up front, wouldn't you be tempted?

new developments

As of April 2007 every new public sector home built must meet the Department for Communities and Local Government (CLG) Code for Sustainable Homes. This code is an updated version of the BRE Ecohomes rating introduced in 2000 that aims to encourage sustainable building programmes and also arm the home buyer with information on the energy efficiency of any new home. Every new home is given a star rating out of six, with six being the most efficient. A development awarded a six will be a super-insulated carbon-neutral home, while a three-star will offer carbon dioxide savings of around 25 per cent compared to general housing stock. The minimum standards set for the code are higher than current building regulations, which should hopefully force builders to aim higher so a rating of just one star is still above basic building regulations. The aim, however, is to set minimum standards closer to three stars than one.

what you can expect from your new-build?

According to the Government Code for Sustainable Homes this is what you get for your star rating:

one star *

10 per cent more efficient than required by 2006 building regulations.
Improved energy efficiency of walls, windows and roof.
Uses no more than 120 litres of water per person per day.
Some materials used being D grade BRE Green Guide (A+ to G).
Provision for waste/recycling storage.
Extras including drying areas, cycle storage, onsite recycling facilities, home office space.

three star ***

25 per cent more efficient than required by 2006 building regulations.
Uses more insulation, condensing boilers, better quality glass, plus possible use of solar thermal or biomass water heating technology. Uses no more than 105 litres of water per person per day thanks to low-flush toilets, flow-restricting taps, energy-efficient dishwashers. Some materials used being D grade BRE Green guide (A+ – G). Extras including energy-efficient lighting, environmentally friendly materials, cycle storage, recycling storage and facilities, home office.

six star ******

The building will be zero carbon from all energy use in the home and will exceed the current building regulations to the point where a negative carbon footprint might be possible – selling power back to the grid for instance.Uses a combination of technologies to meet all the building's energy demands, including solar thermal, micro CHP (combined heat and power), biomass boilers, wind turbines and the like.
Using no more than 80 litres of water per person per day thanks to low-flush toilets, flow-restricting taps, energy-efficient dishwashers. Extras including environmentally friendly materials, minimising building waste, cycle storage, access to water butts, improving solar gain, access to recycling facilities and incorporating energy-efficient lighting and appliances throughout.

five of the UK's best carbon-friendly developments

1: Hockerton Housing Project, Nottinghamshire (www.hockertonhousingproject.org.uk)

This small development is the UK's first earth-sheltered, self-sufficient ecological housing project. The residents of the five houses generate their own low-carbon energy thanks to solar panels and a 6 kW wind turbine. They harvest their own water (both drinking and grey) and recycle waste materials, all of which causes no pollution or carbon dioxide emissions, making this a truly zero-carbon project. The buildings were designed to make the most of the sun's energy, they are highly insulated (helped by tonnes of earth on the green roof, and triple-glazed argon-filled windows) – so much so that there is no traditional central heating. As a result the houses use one-tenth of the energy of a similar-sized conventional house (approximately 8-10 kWh/day).

Whereas other eco-developments often encourage community spirit, at Hockerton it's part of the lease agreement, with each household required to do a minimum of 300 hours of work maintaining the success of the project.

2: The Wintles – Bishop's Castle, Shropshire (www.livingvillages.com)

Where BedZed (see p.298) is a sleek example of high-density modern living, this collection of 40 two- to six-bedroom houses harks back to a more traditional, relaxed village feel. All the properties have been built using non-polluting, renewable, local or reclaimed materials with a high thermal mass to keep in the heat. Some incorporate solar PV panels, they're plumbed for grey water recycling, have a flue ready for a wood burning stove, all feature highly insulated triple glazing, solar hot water heating and a communal LPG boiler provides warmth to the under-floor heating, meaning electricity bills can be as little as £3 a month.

jargon buster

Grey water recycling

Running a bath, taking a shower or just running taps – they all produce 'grey water' that's fine for using in the garden, for flushing the loo and even in the washing machine. Collecting grey water is a simple energy-saving thing to do. New eco-developments should include grey water recycling in the planning. For DIY you'll need a diverter valve that lets you to choose between sending waste water down the drain or to a storage tank.

WOODEN SAY NO:

Low-carbon living in a relaxed village atmosphere at The Wintles in Shropshire.

The houses all line a classic communal village green area where no cars are allowed, making it a safe social space. There's an on-site car club, a nursery and established local food deliveries. It's a short stroll to the centre of Bishop's Castle. It's not a cheap option for Shropshire, with prices ranging from £295,000 to £500,000+, but when they come on the market they don't stay there for long.

3: Great Bow Yard, Somerset (www.swecohomes.co.uk)

On a peaceful canal-side location in rural Langport, Somerset, sit 12 new houses, each with superior levels of insulation. They're all positioned to make the most of natural daylight and have high-efficiency condensing boilers and either solar hot water heating or solar electricity to keep down carbon emissions and energy bills. There are plans to convert a Grade II listed warehouse to similar high standards and include community facilities, with the possibility of a wind turbine and solar P.V system at a later date. To find out what it's like to live in one of these houses read John and Elizabeth's story on the next pages.

4: BedZed, (Beddington Zero Energy Development) Sutton, South East London (www.zedfactory.com)

The UK's largest carbon-neutral housing estate was initiated by BioRegional and developed by the Peabody Trust

in partnership with BioRegional, and designed with Bill Dunster Architects. Completed in 2002, it was built using renewable locally sourced (within a 35-mile radius) materials and wood certified by the Forest Stewardship Council (FSC). The buildings incorporate south-facing terraces, roof gardens, sunlight, solar energy and high levels of insulation, not to mention on-site sewage treatment and grey water recycling. The homes require very little energy to heat, and are all fitted with efficient lighting and appliances. Where this development really stood out, however, is through the use of a combined heat and power (CHP) plant that supplies warmth and electricity to residents using a boiler powered by wood off-cuts that would otherwise end up in landfill. The CHP plant had some teething problems, mainly because it was a prototype: the supplier of the system went into administration before they were able to solve the technical problems, and the project was at the time of writing taking some power from the

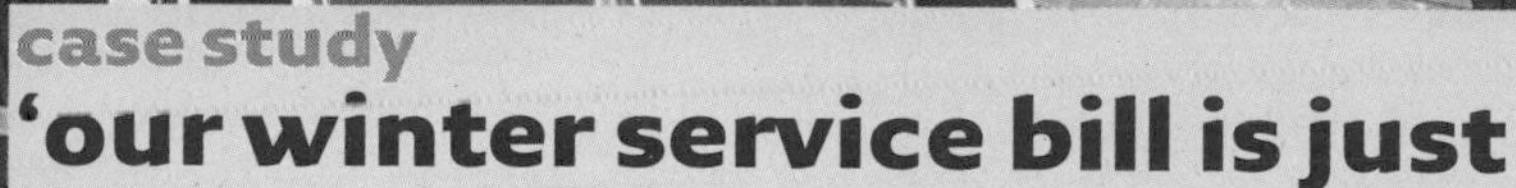

case study

'our winter service bill is just £10 a month'

John and Elizabeth White had always lived a green life, and when they retired they wanted to downsize and cut living costs while also keeping their eco-credentials intact. They bought a three-bedroom home at the South West Eco-Homes Great Bow Yard development in Langport, Somerset. It's a highly insulated terraced building with a modern, efficient living space part-powered by a solar PV system, condensing boiler and enormous highly insulated windows. 'This building has exceeded almost all my expectations,' explains John. 'Because we're in during the day we can really make the most of the sun. Our solar panels provide us with half the electricity we need just because we're in when it's out. We get up, put the slow cooker on which will bubble away slowly until dinner time, use the washing

machine and work in the office – all from just a small solar set-up. It's fantastic that we get to use the power as it's produced rather than importing from the grid all the time. We also rarely use the central heating thanks to the highly insulated walls (they use recycled paper called Warmcel). And because of the east-west positioning we're always kept warm by the sun, not to mention having an upside-down house with the living room and kitchen (where we spend most our time) upstairs. As heat rises, even first thing in the morning, it's toasty and warm. I think it's crazy that such simple and effective building ideas are not used more. We pay about £10 a month in winter for heat and hot water just because some thought went into the design.'

As for a sense of community John couldn't be happier. 'We're remote enough to feel connected to nature but we've got an organic farm about twenty minutes' walk away that sells seasonal vegetables and there's a diverse group of people (from new-born to 82) who enjoy the communal gardens and have a more open mind set, so there's less hiding behind our own front doors and much more interaction.'

grid. A car club discourages unnecessary driving, there are on-site charging points for electric cars and the used of bikes and public transport are encouraged – it's very near two railway stations, two bus routes and a tram link. Best of all, houses prices started from just £102,500 when they were released. To book a tour, visit www.bioregional.com.

5: Looking to the future: Gallions Park, East London

Gallions Park, on a brownfield site in East London, is to become the first zero-carbon development in the UK. Backed by the London Development Agency (LDA), the plan is to build 233 residential units that use 40 per cent less energy than required by 2006 building regulations. Electricity will be generated on-site using a combined heat and power plant that uses sustainable biomass such as woodchip for fuel. The site will be developed by Crest Nicholson BioRegional Quintain LLP and will include building-mounted wind turbines and solar photovoltaic panels. It will also concentrate on providing essential community buildings, recycling and composting, and residents' car clubs. Local, recycled and reclaimed construction materials will be used to help cut carbon miles.

say goodbye to stamp duty

In March 2007 Gordon Brown scrapped Stamp Duty on all new zero-carbon homes until 2012, up to a value of £500,000. That's a potential saving of up to £15,000 (3 per cent) on the total price you pay for a property. It's just a shame that there are only a handful of truly zero-carbon developments available yet, and most of those are unlikely to come up for sale. But hopefully the move will encourage house builders (large and small) to invest in the latest low-carbon building methods.

zero carbon homes for all

Learning the lessons of the BedZED eco-village in south London, BioRegional Quintain Ltd is building more than 1,400 homes from Brighton to Middlesbrough that are highly energy efficient and run on renewable energy.

Riverside One (pictured) is the largest zero-carbon scheme in the UK with a planned total of 750 homes, offices, hotels and leisure facilities. Next to Middlesbrough Football Club, the site is being regenerated using some of the UK's leading architects including Will Alsop, Studio Egret West, FAT, Grimshaw and Feilden Clegg Bradley Studios.

BioRegional Quintain's One Brighton includes 172 apartments and community facilities next to Brighton station. It uses natural clay blocks and wood fibre insulation which mean the homes need very little heating. A wood-fired boiler will provide zero carbon heating and hot water – about half of the total energy demand for the homes. Photovoltaic panels and roof-mounted turbines provide some of the electricity. The rest of the electricity will come from large wind turbines offsite, bought in bulk to supply competitively to residents. One Brighton is being built in partnership with leading UK housebuilder, Crest Nicholson PLC and will sell at prices comparable to others in Brighton. A third of the homes will be low cost rent and shared ownership. The development not only aims for energy efficiency, it is designed to support a healthy and green lifestyle, with roof top mini-allotments, sky gardens, rainwater recycling, a car club and on-site composting. Construction started in September 2007 with on-site generators running on waste cooking oil – reducing the carbon emissions associated with the building works.

'If everyone consumed as much as we do in the UK, we would need three planets to support us. With greenhouse gas emissions contributing to about half of our ecological footprint, we need to create places where it is easy, attractive and affordable for people to adopt a one-planet lifestyle.'

Pooran Desai, OBE, Sustainability Director of BioRegional Quintain
www.bioregional-quintain.com

building your own home

In 2005 around 25,000 people chose to build their own homes. It's an area where being as efficient, low-impact and sustainable as possible has always been a top priority. Free of the corporate shackles imposed on big property developers, self-builders have the freedom to use whatever materials they choose.

build your own eco-home at night school

From the makers of BedZed comes RuralZed, a Cornwall-based housing project that hopes to train up wannabe self-builders to construct their own £75,000 eco-home, complete with recycled railway sleepers, solar panels and highly insulated timber-framed construction. Training courses cost £5,000 and run for ten weeks (minimum one day a week) and they'll teach you everything you need to know to construct a locally sourced timber-framed house with solar panels, recycled railway sleepers and advanced ventilation.

embodied energy

Add up all the processes involved in the building of your home, from extracting raw materials through to transporting them and you'll soon see just how energy-intensive some methods are.

For example, a basic timber-frame construction would have an embodied energy of 188 MJ/m^2 (Mega joules per square metre), but the same construction using traditional brick and block building would be closer to 906 MJ/m^2 – nearly five times more energy. The embodied energy of timber-based building products is up to 60 per cent lower than masonry, making it by far the least polluting building material to choose for any construction or renovation project.

hemp and tyres: the latest green building methods explained

Looking to renovate, rebuild or start a project from scratch? Here are some of the greenest building methods available.

Earth constructions:
A third of humanity still lives in earth-based buildings, and it only takes a look towards some of the world's poorest countries to see why these methods (usually of stacking and compressing earth to form 'cob' walls) are some of the most efficient, least intrusive construction methods available. Cob is made from the soil – not organic top-soil but the layer beneath that's rich in clay. This is mixed with water, straw and sand and, when formed into walls (usually on top of a small stone wall to prevent damp rising into the cob), it produces a hard-wearing, long-lasting structure that, once dried, can be rendered with natural lime wash or lime plaster and painted like a conventional wall. It's such a low-impact building method that for all but the biggest builds the only tools you need are yourself, a tarpaulin, wheelbarrow and a spade.

Straw bales:
Quick, cheap and easy, with superb insulating qualities, straw bales can be used in a number of ways – all of which are highly insulating and very low impact. Use them to fill in timber frames, or as building blocks, either way the advantages over modern intensive building methods are enormous. If used to replace brick and block construction it can save you around £10,000 – if you're building a three-bedroomed house – and you'll enjoy a huge reduction (up to 75 per cent) in heating costs once the house is occupied.

Bio-bricks:
If you want to stick to the traditional brick and block building method you could try using Bio Brics. This new system uses special interlocking bricks that have a honeycomb system of cavities within them meaning there's no need for cavity wall insulation. The bricks are produced in kilns that run on 40 per cent sawdust and the system does not use any plastic or chemicals that give off harmful fumes. For more details visit www.limefirms.co.uk.

Structurally insulated panels:
Commonly known as SIPS, this is fast becoming the self-build method of choice. Timber walls, complete with insulation, windows and even wiring, are constructed off-site in a factory and delivered on several large trucks before being bolted together on-site. It's a quick and flexible building method that uses up to 50 per cent less raw timber than a traditional timber building, and takes five times less energy – not to mention considerably less waste – to produce than concrete but still provides a highly insulated construction. Many flat-pack, or kit house, firms offer

WONDERWALLS:
Tyres, bottles and cans being re-used for building – a great way to reduce embodied energy, and deal with waste

this building method, including Potton (www.potton.co.uk) but for general advice on timber building visit the UK Timber Framed Association (www.timberframe.org).

Hemp building:

Mixed with lime, hemp is a very useful building material (often known as Hempcrete) that can be used as insulation in lofts or between floors. It can also be used to build walls, especially with timber-framed constructions such as post and beam. For more details visit www.lhoist.co.uk/tradical/hemp-lime.html

Recycled waste:

With a little imagination and careful planning, much everyday waste destined for landfill can be reclaimed and re-used in green-building. A good example is the use of old tyres in the Earthship Brighton development. This groundbreaking build uses more than 2,000 old tyres instead of concrete for many of the walls and foundations, greatly reducing the embodied energy of the building, and slashing its carbon footprint. Don't worry, they don't lead to bouncy foundations: each tyre is packed with compacted earth and other recycled items such as glass bottles.

carbon-neutral kit house

Self-build specialist Potton is the first UK company to introduce a kit home that means self-builders can build a net zero-carbon home with very low annual running costs.
The Lighthouse range of homes meets the government's six star for sustainable building, thanks to a combination of SIPs (Structural Insulated Panels) that bring thermal insulation and air-tightness, low water, sanitary ware and appliances, rainwater harvesting and grey water recycling, as well as renewable energy technologies, including a biomass boiler, PV solar panels and a solar-thermal water heater. The house will cost from £175,000 excluding land costs. For more information visit www.lighthousebypotton.co.uk.

www.greenmoves.com

They're cheap to run and comfortable places to live, so it's little surprise that eco-homes don't come on the market very often. When they do crop up the chances are you'll find them advertised on www.greenmoves.com, the UK's only totally green property portal. With sales and rentals listed, each property has to meet a set of standards based on BRE Eco Home ratings if they're new builds and Home Energy Ratings/Energy Performance Certificates if they're older properties. Each house is given a mark out of 120, and only those with ratings over 90 will be accepted.

MICROGENERATION – power to the people

The ultimate good life: produce your electricity from the wind, hot water from the sun and heat from the ground to make your home or business carbon-neutral (or even carbon-positive), liberated from fossil fuels and spiralling energy costs. In 2005, total UK carbon dioxide emissions were 556 million tonnes, of which around 37 per cent were created by generating electricity, with a further 15 per cent coming from fossil fuel use in the home.

There's been enormous media noise about micro wind turbines and solar power and some outlandish claims over their effectiveness. But they do work and should be considered part of the answer to combating climate change – as well as managing energy bills. There has never been a greater range of options available and, used well, they can make a massive dent in your carbon footprint. Here are the best and what you can expect for your money.

good foundations

Successful micro-generation projects all have one thing in common – they are all used in addition to (often simpler) energy-saving measures such as ensuring that a home is as well insulated as possible, and switching to energy-saving light bulbs. It's only worth trying to produce energy when you've saved as much of it as possible at home, and it's only then that solar power, wind turbines and heat pumps come into their own.

heating

solar hot water

What is it?
A series of tubes mounted on south-facing roofs that use the heat of the sun to warm a solution of water and anti-freeze. This is pumped through a coil in the hot water tank, heating it up. Once the heat has been transferred to the water tank the solution is pumped back to the roof for reheating. Temperatures of 60°C are easily achievable – more than hot enough for baths and showers. Solar water heaters can't heat our homes – they only supply water for baths and showers, but they can be used to heat swimming pools.

How much energy will it produce?
Even with grey, cloudy British skies a small-scale system can generate 50 per cent of a house's hot water needs, and almost 100 per cent over the summer months. The systems rely on radiation rather than direct sun. It is quite possible that during the summer when your central heating isn't needed you will not need to turn your boiler on at all.

Is my house suitable?
You'll need 2-5 square metres of south-west-to-south-east-facing roof that receives as little shade as possible, plus room inside for a water tank (around 300 litres) if you don't already have one (they are not compatible with combination boilers). The tubes need to be mounted at 35° from horizontal to get the most from the sun's rays.

How much will it cost?
Expect to pay £2,000-£3000 for a flat plate (the cheapest with around 30 per cent efficiency) installation and £3,500-£4,000 for a more efficient (up to 70 per cent) evacuated tube system. Grants of up to 50 per cent have been available in the past. Available now on the high street – B&Q sell hot-water heaters for £1,498 (without installation).

How much cash will I save?
As systems can last up to 25 years with minimal maintenance you'll make substantial savings in the long run. Per year, if the conditions are right, you could save £50 on your heating bills.

How much carbon will I save?
The Energy Saving Trust estimates savings of 350 kg of carbon dioxide per year. Over a 20-year period this would imply a saving of 7 tonnes of carbon dioxide. In the summer if you're using electricity in the kitchen and avoid the central heating you won't need any gas at all.

What should I use?
Talk to the experts, get a full site evaluation and as with any building work get a series of quotes first. Ideally choose the more efficient evacuated tube designs. Visit the Solar Trade Association (www.solartradeassociation.org.uk) for a list of installers.

case study
'even on a rainy day we still reap the benefits'

Martyn Williams, from East London installed solar hot water panels to his Grade II Georgian terrace.

'Once we'd moved in we took stock and decided we'd like to replace the old boiler with a combined heat and power system, and mount solar thermal collectors to produce our hot water on the south facing bay window roof. We talked to a few neighbours about what they would think about this and they were mostly very positive. We were worried about the grade II status, and the fact that we had a National Trust property overlooking us, but to our relief we got a lot of support and no opposition at all to our plans. We submitted our Listed Building application in January 2007 and got the go ahead thirteen weeks later.

'The total cost of boiler and panels will be just under £8,000. By now we had given up on sourcing a micro CHP plant – the companies who have been promising them still had not come forward with one. Instead we chose the most efficient condensing boiler we could find. The panels and new hot water tank were fixed in place by the end of June,

and from day one they've worked beautifully.

'Even though the weather was atrocious, even for a British summer, we barely used the boiler – the sun has provided us with pretty much everything we need. On the best day we have had, a relatively sunny day but still with some big fluffy clouds around, we had a complete 210-litre tank of water at 65°C by lunch time – put on two loads of washing to take some off but we were fully reheated by 3.30 pm, at which point the pump automatically shut down (the tank is limited to 65° to avoid scalding) while the panels went on to reach over 110°C.

'I'm extremely pleased with the panels, as even on the really dark rainy days they have heated the entire tank to mid 35-40°C – a good 15 to 20 degrees higher than the water would be if the tank filled without heating. This difference means the boiler works half as hard to get the water up to a useable 55°C. Compared to the solar electric panels on my last house this is even more satisfying because running a nice hot bath that's cost nothing and is carbon-free is so much better than just flicking a switch.'

heat pumps

What are they?

You already have a heat pump in your house: a compressor and hot coil give off heat and are mounted at the back of your fridge, while the evaporator and cold coil absorb heat inside the fridge. In the same way a heat pump absorbs heat from outside and gives it off to your heating system. Heat pumps work best with a relatively small temperature difference, when they can put three times as much heat energy into your home as they use in electricity. If your only viable source of heat is electricity, then a heat pump will cut your running costs and carbon emissions by up to two thirds.

The cheapest form of heat pump extracts heat from the outside air, but when the air is colder the heat pump gets less efficient, so most modern ones extract heat from the ground, where below a metre or two down the temperature is roughly constant year round. These are known as Ground Source Heat Pumps (GSHPs). Heat pumps are often used with underfloor heating, although they can also be used with very large radiators. Either way the house must be well insulated.

The downside is the cost of installing the heat exchange pipes that absorb heat from the ground. If you have a large garden or field which you don't mind digging up, a contractor can dig trenches for the heat exchange pipes; if you don't they can drill a borehole for the pipes. Alternatively a stream or lake is an ideal source to extract heat from.

How much will it cost?

Heat pumps are not cheap. With 7-8 kW systems (average installation size) costing about £1,000 per kW of energy needed you can expect to pay £6,500-£11,000. Grants have been available to help with the costs.

How much cash will I save?

Compared to electricity or oil-fired heating your running costs will be a good bit lower.

How much carbon will I save?

Compared to electric heating (off peak or standard rate) you will save about two thirds. Compared to oil heating you will save some carbon, and this saving will get better over time as the UK electricity system becomes less carbon intensive.

What should I use?

Talk to the experts and request a site survey to see if your home is suitable. Do you have enough land, is your home insulated well enough to only require low temperatures to keep it warm? For a list of installers visit www.lowcarbonbuildings.org.uk/micro.

electricity

wind turbines

What are they?

There are dozens of different turbines available, including very low power kits suitable for boats and sheds. For serious power generation, however, you have two choices: the micro, typically under 3.5 kW power with a blade size of 4 metre diameter, and small scale, under 50 kW with a diameter of less than 16 metres. The turbines rotate in the wind turning a generator that produces electricity which is then either stored in batteries, or fed through a transformer to be used with a standard 240-volt supply or with larger systems fed straight into the mains electricity grid. For an average household to meet all their electricity needs from a small wind turbine a system of between 2.5 kW and 6 kW would be needed, but this greatly depends on the wind speed available and is unlikely to be possible in built-up urban areas. And don't forget the mast: the higher off the ground the turbine can be the more consistent and reliable the wind will be.

Can I do it myself?

Small DIY kits are available and the Centre for Alternative Technology (www.cat.org.uk) even runs courses on building your own turbine, but these are generally only smaller devices offering top-up power or used for educational purposes. If you're looking for a bigger, more powerful system contact the professionals via the British Wind Energy Association (www.bwea.com).

How much will it cost?

A micro 1.1 metre wind turbine with a peak output of 500 W costs around £700 excluding installation. A larger 2.7 metre, 1 kW wind turbine connected to the grid will cost from £3,300 while a turbine that would be powerful enough to meet the needs of an extremely insulated house, 3.5 metre diameter and 2.5 kW, would cost around £5,500.

How much cash will I save?

It is unrealistic to think that a micro/small wind turbine on its own will meet all your electricity needs, but it can be a sound investment. A small turbine costing just a few hundred pounds can charge a battery that in turn can power garden lights, sheds, charge mobile phones etc, while a larger turbine located with a reliable wind supply could contribute significantly to your overall energy needs.

How much carbon will I save?

The Energy Saving Trust estimates that a 2.5 kW turbine running at maximum capacity can save 1-2 tonnes of carbon dioxide per year while a larger 6 kW system can save 2.5-5 tonnes of carbon dioxide per year. The 1kW Windsave micro-turbine famously sold in B&Q has the potential to save between 249 kg and 377 kg carbon dioxide if working at 5.8-7 metres per second wind speed (17-27 per cent saving in electricity costs). The UK's average wind speed is 6 metres per second (average over a year); Downing Street measures just 4.3 metres per second.

Where should I start?

Check your average wind speed by visiting the British Wind Energy Association website www.bwea.com, type in your postcode and you can easily forecast how suitable your site is. For your turbine to start producing electricity you'd need a constant wind speed of 3-4 m/s but most products achieve their maximum output with wind of 10-12 m/s. Don't take this figure as a green light to order a wind turbine, however. You'll need a complete site survey from the manufacturer to determine possible locations and output. If you're in a low-lying built-up area with patchy wind speeds a turbine might well be a very expensive way to power just a couple of light bulbs. Efficient micro-turbines are not quite with us yet, but as demand increases and technology improves things will slowly get better.

If you are lucky enough to have the land and the constant wind speed you're likely to make a great big dent in your carbon footprint if you install the correct size turbine for your needs.

Remember, no matter where you want to site your turbine, you'll need to get planning permission. In a bid to encourage more people to install micro renewables the government planned (from June 2006) to relax the planning restrictions on domestic turbines. You'll still need to apply in the same way – visit www.planningni.gov.uk for more details – but the process should be smoother and with less obstructions.

Colour House turbine

London's first ground-based wind turbine, at The Colour House Children's Theatre in Merton, is designed for urban areas where winds tend to be more turbulent. The helical (twisted) design is quieter, and vibrates less than traditional designs and its compact size means it can fit into tight spaces. It is hoped the Colour House turbine will provide all the theatre's electricity needs. Purchasing and installing cost around £33,000 and with a constant wind speed of 5.9 metres per second could pay for itself in 18 years. www.quietrevolution.co.uk

case study

you become very attached

Internationally renowned artist Kurt Jackson lives and works in a carbon-neutral home and workshop in Cornwall. The Friends of the Earth supporter has installed a wind turbine, heat pump and solar PV system to meet his electricity needs, using a heat pump for heating and hot water.

'We employed a green consultant who looked at our energy needs and suggested solar, wind and geothermal technologies would work best. Getting the turbine approved by the local authority was a bit of a palaver as ours was the first in the district. The only objection came from the local wing of the National Trust, but after negotiations with their head office it was thankfully approved. That said, it did take quite a bit of organising including two council meetings and a site visit before we got everything sorted. We chose a 6 kW turbine that produced 14,000 kW of electricity in 2006 – more than enough to power the house. Our 5 kW solar panels help power my workshop, which makes us entirely self-sufficient in energy. We have a deal with the national grid, so if we produce more than we need we sell it to them, and if we have a shortfall we buy from the grid.

'Choosing to invest in renewable technology has opened my eyes to my own carbon footprint, my energy use and exactly how much power we consume. You become very attached and involved and it's incredibly satisfying once it's up and running – plus our attitude to the weather has changed completely. There's no such thing as bad weather anymore.'

solar power (photovoltaics)

What is it?

Using the power of the sun to generate electricity. After all, the sun produces enough energy in one hour to meet the world's needs for two years – with power like that it would be a shame not to exploit it. Solar photovoltaic (PV) systems convert the sun's energy by creating an electric field across layers of silicon, causing electricity to flow. The silicon strips are housed in casings that need to be mounted at 30-45 degrees facing south to make the most of the sun, which is why rooftop installations are the most common. The greater the light, the greater the flow of electricity, although these systems will generate power, albeit at a lower level, even on cloudy days. The panels perform at the level of the one receiving the least light so even a small area of shade can reduce the output of the system.

How much will it cost?

This greatly depends on how much electricity you want to produce, but as a general rule a 2 kWp (peak output) system large enough to provide half the electricity of a three-bed house will set you back around £12,000. (Fifty per cent grants have been available). Because electricity is still extremely cheap compared to the amount you'd need to invest in solar, a system won't pay for itself for many years. Remember, you're investing in a low-carbon life, not a power station.

How much cash will I save?

With a 2 kWp system you'll save around £150 per year on your electricity bill and any electricity your home doesn't use can be sold back to the grid.

How much carbon will I save?

If a 2 kWp system produces 1,500 kWh of electricity per year (solar systems rarely work to their maximum capacity) you can expect to save 645 kg of carbon dioxide per year. This increases the more panels you have and the sunnier your site is; so a family home with a large (4 kWp) system could save more than 2 tonnes of carbon dioxide per year.

What materials should I use?

Your choice of panel will depend on the size of roof you have (ideally you'll need 10 square metres of unshaded roof facing roughly south), how sturdy it is (panels can weigh up to 40 kg per square metre), and what finished look you're hoping for. Standard installations are usually mounted on rails directly onto existing tiles, but panels that can be used instead of roof tiles such as the C21e roof tile panel from Solar Century (www.solarcentury.co.uk) are increasing in popularity. You'll need a full site survey before any system can be installed and this will help to determine the best panels for your property.

case study

'Now we're producing our own electricity we're much more careful how we use it'

The Merrill family from Somerset installed 36 solar tiles to their home, covering 15 square metres of roof space at a cost of £12,960 (plus £1,500 installation and including a £2,500 grant). They chose flush-fitting tiles from Solar Century that blend with their existing roof tiles. 'We wanted to effectively reduce our carbon footprint and electricity bill but didn't want to alter the look of our home, so these tiles were perfect,' explains David Merrill. 'We have a 2 kW system that in the first six months saved us 1 tonne of carbon dioxide. And because we're now more careful with the energy we use we've managed to cut our electricity bill from £150 a quarter, to just £50 – that's a saving of £400 a year.'

from little acorns...

It doesn't have to cost thousands to enjoy the benefits of renewable energy. Many garden lights are now combining low energy LED technology and small solar panels to great effect and there are many small kits available, perfect for lighting your shed or caravan, that cost around £80 (www.greenshop.co.uk). Mini chargers such as the stylish Solio (www.solio.com) can help top up your phone or MP3 player for free. Mobile phone manufacturer Motorola has recently submitted a patent for a solar-powered mobile phone screen, so it might not be too long before we see self-powered gadgets.

case study
'free hot water – how rewarding is that?'

'I have made many, many small changes in my attempts to cut carbon, most of which have been easy and rewarding. Yes, it is a pain sometimes to build in a trip to a bottle bank while out and about but it is so satisfying to smash glass for a good cause.'

'It's not cheap to put in solar power but how rewarding to get free hot water. We have to think ahead to get to the farmers' market or the local organic shop without increasing our carbon miles, but the food definitely tastes much better. It takes more time to work out how to dispose of anything, eBay, Freecycle, recycle, offer to a friend etc, but it feels great seeing something leave the house that is not going into a tip. One of the most painful changes was getting back on my bike after 20 years. I have now invested in a bigger saddle.'

Sue Davis, Green Champion and Friends of the Earth supporter.

micro CHP – the next big thing?

Hailed by many as the future for domestic heat and energy production micro CHP (Combined Heat and Power) is a revolutionary energy production method that generates both electricity and uses the waste heat – it's effectively like having a mini power station at home. Electricity generation at a traditional fossil fuel burning power station is less than 50 per cent efficient, but electricity generated through a micro CHP boiler will be 80–90 per cent efficient, saving money and fossil fuels. Large-scale combined heat and power plants such as the town centre development in Woking, supplying heat and power to shops, services and houses are already proving successful, greatly reducing heating costs and cutting carbon dioxide emissions. The scale is gradually getting smaller, with schools, community centres and office blocks benefiting from the savings (of both carbon dioxide and running costs). The School of the Built Environment at Nottingham University installed two Dachs mini-CHP units and so far they have been running at 80 per cent efficiency, and saving over £11,500 in heating and electricity costs.

With this in mind it shouldn't be long before domestic CHP systems are being installed in people's homes just like a regular boiler. It's predicted that these boilers could save a quarter of a home's or business's carbon dioxide emissions, or around £150 a year, and further down the line the Energy Saving Trust hope micro-CHP units could provide 25 per cent of UK electricity, which would greatly reduce the need for large, inefficient power stations.

whispergen

Designed by Powergen, the Whispergen is hoped to be one of the first domestic micro CHP boilers widely available. The developers predict the unit will reduce household carbon dioxide emissions significantly. They have been testing the units for some time and are expecting it to be ready for sale by 2009.

out and about

chapter 8

It broadens the mind.
Travel greener.

ON THE ROAD

Cars are responsible for around 13 per cent of UK carbon emissions and road traffic and emissions have risen by 10 per cent since 1990. The average petrol-driven car does 9,000 miles a year and emits 3 tonnes of carbon dioxide (its diesel equivalent emits 2.9 tonnes). Using the car is an unavoidable reality for most of us, but that doesn't mean we can't still cut our emissions. All it takes is a few simple changes. Changing the way you drive, joining a car club, or considering a more efficient car the next time you buy can all cut carbon dioxide emissions significantly. And there's nothing to stop you from sometimes taking trains and buses, and walking or cycling. Even if you only do this occasionally it will still make a huge difference.

becoming a greener driver

Most of us can cut the carbon our cars belch out just by changing the way we drive. These easy tips make driving greener, safer and more comfortable for everyone:

1. Slow down. You can improve fuel economy by 25-35 per cent by travelling at 70 mph not 90 mph. If you accelerate and brake gently, especially when starting and stopping, you'll also be able to improve fuel efficiency by up to 30 per cent.
2. Shift to a higher gear at the right time. Shift up at 2,500 rpm for petrol cars and 2,000 rpm for diesel cars: a vehicle traveling at 37 mph in third gear uses 25 per cent more fuel than it would at the same speed in fifth gear.
3. Buy a map. According to the AA, getting lost wastes hundreds of thousands of gallons of fuel a year. Use one of the many online route planners to find the best directions.

carbon dioxide emissions

Travelling is frequently cited as an important factor in climate change. But how does flying compare, say, with driving a car? And how much difference does taking public transport really make?

emissions from UK transport (millions tonnes CO_2 per year)

Total emissions from domestic transport:	**129 (17%)**
Passenger cars:	69.9 (12.6%)
Light duty vehicles:	16.8 (3 %)
Buses:	3.6 (0.6%)
Heavy goods vehicles:	28.6 (5.2%)
Mopeds and motorcycles:	0.4 (0.1 %)
Railways:	2 (0.4%)
Other:	7.7 (1.4%)
International aviation	**35**

carbon emissions per kilometre per person (g CO_2 per km)

Means of transport	**Emissions of CO_2 per km**
Car (the most fuel efficient available)	90-120*
Car (average models)	160-200*
Car (large models, SUVs etc)	200-500*
Rail (normal suburban)	40-130
Rail (high speed, few stops)	80-165
London Underground	56
Bus	104
Tram	65
Coach	28
Air (short haul)	330-460
Air (long haul)	210-330

* per vehicle - share the journey and emissions per person go down

embodied energy in car production

Car manufacturers are often very happy to trumpet their vehicles' green credentials, but you never hear them talking very much about the amount of energy used to produce them. In fact, it's, notoriously difficult to get these energy figures out of manufacturers – partly because it would give away how much the car actually costs them to produce, but also because it's difficult to factor in the lifecycle of every nut. bolt and screw. It has been estimated that 25 tonnes of waste are created in the production of each car, and that the manufacturing stage is responsible for 10-20 per cent of a car's total energy consumption and emissions.

greener driving route map

Drive less: Do you need to make the journey? If so can you walk, cycle, take public transport or share a lift?

Drive more efficiently: If you need to make that journey, take it easy and drive in the ways recommended to save most fuel. And is your car performing up to scratch?

Choose the greenest car: The next time you buy or rent go for the lowest emissions/highest miles per gallon.

carbon calculator – driving

Log on to www.vcacarfueldata.org.uk – a national database providing figures on car emissions. Look up the make and model of your car and see what the grammes of carbon dioxide-(CO_2)per-kilometre figure is. Then estimate how many kilometres you drive each year (if you work in miles, multiply the figure by 1.6). Multiply the two figures together and divide by 1,000 to get your total carbon dioxide emissions in kilograms.

The following examples are based on an average of 9,000 miles (14,400 km)

		Annual emissions (kg CO_2)
Aston Martin DB9 (the 'James Bond car')	=	6,062 kg CO_2 per year (over 6 tonnes)
Ford Mondeo 1.8 Duratec petrol	=	2,649kg CO_2 per year (2.6 tonnes)
Vauxhall Corsa 1.3CDTi diesel 3 door	=	1,656 kg CO_2per year (1.65 tonnes)
Smart fourtwo coupe petrol	=	1,627 kg CO_2 per year (1.62 tonnes)

4 Avoid using the air-conditioner. Use the air vents to maximise fuel efficiency, but avoid opening the windows, especially at high speed as this can increase the amount of fuel consumed. Ironically at 80 mph it's more fuel- efficient to use the air conditioner than to open the windows. That said, you shouldn't be going at 80 mph, should you?
5 Keep your tyres pumped up. For every 6 psi pressure a tyre is under-inflated fuel consumption rises by 1 per cent.
6 Aim to get as much as you can out of the journey. Doing five jobs in one trip is better for the environment than making five trips.
7 Don't leave the engine ticking over. Switch it off if you're going to be stationary for more than two minutes. Leaving the engine ticking over wastes fuel and wears down the engine. Modern cars don't need to be warmed up.

top tip

Before you think about buying a new, more efficient car ask yourself if you've done what you can to make your existing one as efficient as possible. Have you changed your driving style? Remember, your car could have created up to 25 tonnes of waste during its production. And if you've been on the M1 recently you'll know this country doesn't really need any more cars on its roads – however eco-friendly the model.

how the government could help

Let's face it, we're not all going to give up our cars, but if we became a bit more independent of them we could really dent our carbon footprints. It isn't easy, which is why it's vital that the government gives us lots of reasons to go back to public transport. Here are a few steps they could take towards reducing the number of cars on our roads:

- Introduce mandatory carbon dioxide limits for new cars to double current fuel efficiency by 2020.
- Invest in smarter choices, e.g. travel planning for companies and infrastructure for cycling and walking.
- Improve public transport – more reliable, regular and accessible buses.
- Make streets safer – lower speed limits to 20 mph in towns and 10 mph in residential areas. Prioritise pedestrians over cars in residential areas.

- Spend less public money on building roads and more on improving alternatives to the car.
- Increase the costs involved with running a heavily polluting vehicle, and make it cheaper and more convenient to choose low-emissions vehicles.

buying a greener car

Hybrid, diesel, electric, LPG, biofuel or petrol? The choice is growing all the time. Here's what you need to know about the latest green machines and the alternative fuels that are around.

THE CAR'S THE STAR: Farmer Peter Kindersley charges his G-Wiz electric run-around at night. The car can cover 40 miles before it needs recharging. Peter owns Sheepdrove Organic farm in Berkshire (www.sheepdrove.com).

“If transport is to play its full role in reducing the UK’s climate change emissions, then traffic levels must be reduced. Standards to force manufacturers to build and sell greener cars, and greater incentives for drivers to buy and use these vehicles are absolutely essential, and the UK Government must do all it can to ensure tough action at the European level.

Tony Bosworth, Friends of the Earth's senior transport campaigner

green labelling

The next time you’re looking around a car showroom, home in on the energy labels. Similar to the ones on kitchen appliances, although as yet not compulsory, these labels show at a glance how much carbon the vehicle produces, its fuel consumption, and your likely fuel costs over 12,000 miles. Rated A-G (with A being the most efficient) the most efficient A-rated cars, such as the VW Blue Motion, emit less than 100 g/km carbon dioxide, while the worst will easily produce more than 185 g carbon dioxide/km . The Range Rover Freelander 1.8 petrol creates 248 g carbon dioxide/km . Of course, the worst G-rated emit over 300 g/km..

fact

At least ten per cent of a car’s lifetime carbon dioxide emissions are in its production; 85 per cent in use; and 5 per cent in disposal/waste.

hybrid cars

Hybrids such as the celebrity-endorsed Toyota Prius use a conventional petrol motor (diesel varieties are expected soon) as well as an electric battery that charges as you drive and automatically switches on when the car slows down, making city driving (under 30 mph) much greener. They cost around two-thirds less to run than a petrol vehicle and are exempt from London congestion charging (and will be exempt from other schemes such as those in Manchester); they also have a reduced road tax – but remember, they're not the answer to cutting transport emissions as they're only green when going slowly. The rest of the time they're just regular petrol-hungry gas-guzzlers.

fact

23 per cent of journeys made by car are less than 2 miles long. Park the car, hop, skip or cycle and save yourself 86 kg carbon dioxide per year.

electric cars

With no exhaust emissions, electric cars are currently the greenest and cheapest way to go (assuming your electricity supply comes from renewable sources). Plug them in like a mobile phone, charge up for a few pence and away you go. Suitable only for urban or local driving, electric cars have a typical range of about 40-50 miles and a top speed of about 50 mph, but technology is improving fast. In London and Manchester (other cities are on the way) you'll enjoy free parking permits in almost all boroughs, free access to central car parks, no congestion charge and free road tax. If your electricity comes from non-renewable resources, it has been estimated that electric cars run at the equivalent of about 70 g/km carbon dioxide.

liquefied petroleum gas (LPG)

You'll need to get your existing car adapted at a cost of £1000-£2000, but LPG is more efficient than petrol and has considerably less pollutants than diesel. It's cheaper to fill up with at around half the cost of petrol, although the cost of new LPG cars will be higher – typically, £1,200-£2,000 more than for non-LPG versions. In 2007 there were 223 LPG refuelling stations around the country. Ironically as improvements in petrol efficiency continue the benefits of LPG are waning and previous tax breaks for choosing it are being called into question, although LPG cars are still exempt from the London congestion charge.

biofuels

Produced from plants such as sugar cane, sugar beet, oil seed rape and soya beans, biofuels (bioethanol is a petrol substitute, biodiesel replaces diesel) have been touted as the king of carbon-neutral fuels because all the carbon dioxide emitted during use is absorbed during the plant's lifecycle. Many cars in Brazil are already being powered by sugar cane ethanol. Most petrol or diesel vehicles can already use biofuels, or require only minor modification. The latest Flexi-Fuel vehicles (FFV) can run on any blend of petrol and ethanol, or just standard petrol. Because cars running on pure bioethanol need around 50 per cent more fuel per mile compared to conventional fuel they cost roughly the same to run.

Too good to be true: The cultivation of plants for biofuel is driving massive deforestation in South East Asia (one of the main causes of climate change and the very thing it's trying to prevent) and also increasing competition and conflict over land, driving up food prices which will threaten the lives of some of the world's most vulnerable people. While the technology is undoubtedly greener at point of use the knock-on effect could be disastrous and actually produce more greenhouse gases than it saves.

petrol versus diesel?

Diesel cars are more fuel-efficient than petrol-driven ones, and will on average squeeze out more miles per litre – burning a litre of diesel creates more carbon dioxide than burning a litre of petrol, but the engine efficiency just about makes up for that. However, diesel will create more dirty emissions such as nitrogen oxides and particulates that can affect your health. If you are considering buying a diesel car, choose one with a diesel particulate filter (DPF), as this will reduce these emissions. That said, if you live in an urban area and drive a petrol engine that uses the latest low-sulphur fuel it will be greener than diesel.

hot off the drawing board

Using compressed air instead of traditional fuels the three-seater Minicat from Monteur Development International (MDI) in France could be the next big thing in green motoring. Compressed air is stored in onboard air tanks and used to power the engine with zero emissions and almost twice the range of a typical electric car (which is 60-100 km), up to 300 km. The tanks will be refillable either at home using electricity, or at air-compression filling stations. According to the company's website it will cost less than 1 euro every 100 km, ten times less than petrol-powered cars.

car sharing

If you live down the road from someone you work with and you both drive, sharing a lift instantly halves your petrol costs and emissions, not to mention the wear and tear on the cars. If you can persuade your neighbour to give you a lift to the shops, you'll halve your costs again and if you can alternate the school run between parents and you squeeze in four kids the savings will be even greater. If you've managed to do all this, you're on the way to setting up a car club.

If you've had no luck with the neighbours, try these journey matchmaking websites. There are now more than 165,000 members of LiftShare.com saving an estimated 5,800 tonnes of carbon dioxide every year.

www.liftshare.org
www.carshare.com
www.shareajourney.com

top five carbon-friendly cars

1 **G-Wiz (electric) two seater**
carbon: 0 g/km (assuming you recharge using a renewable supply)
mpg : n/a
range: 40 miles before recharging
price: From £7,299

2 **Volkswagen BlueMotion**
carbon: 99 g/km
mpg: 74.3
price: £11,995

3 **Mini Cooper D hatchback with particle filter**
carbon: 104 g/km
mpg: 72
price: £14,000

4 **Toyota Prius petrol hybrid four-door family car**
carbon: 104 g/km
mpg: 65.7
price: £17,000 - £20,000

5 **Citroen C2 1.4 diesel**
carbon: 107 g/km
mpg: 61.4
price: £7,995 - £10,795

buying second hand

Because green cars are relatively new, and still a small percentage of the car market, tracking down a second-hand green motor might be difficult. If you're looking for a second-hand car check out www.whatcar.com for detailed lists of every car make and model including the carbon dioxide emissions so you can be sure you're choosing the least polluting vehicle.

car clubs

If you don't want the expense of owning and looking after your own car, but need one every now and again, it might be worth joining a car club. A car club provides its members with quick and easy access to a car, usually left parked in designated bays within the local area, that can be unlocked with a smart card. The cars can be booked online for an hour, day, week or month, making it far more affordable than standard car hire, not to mention a more environmentally sensible way to drive.

For more details and an interactive map of all car clubs in the UK visit www.carplus.org.uk.

case study

'I'll never buy a normal car again'

John and Carol Lucas have three kids and couldn't do without a family car so they chose a Toyota Prius in an effort to cut their carbon footprints.

Has owning a hybrid made you more conscious of driving?

'Absolutely. I try and drive as little as possible, we always try to walk to local shops and parks, and get on our bikes more. I must confess to feeling a slight sense of smugness when silently passing parents on the school run stepping out of their huge 4WDs.'

Does it live up to your expectations?

'It probably exceeds them because it's very well made, I think it looks pretty streamlined and it's a very safe car to drive. It's enabled a busy five-person family to manage surprisingly easily, considering we now drive for our holidays in Britain. It's extremely satisfying drifting silently around the Lake District knowing that, for most of the time at least, we're not contributing to its destruction.'

motorbikes

Two wheels are almost always going to be better than four as they use less fuel, which creates less carbon dioxide, take less room on the roads and are easy to manoeuvre. But most are still fossil-fuel users and for urban journeys of just a few miles, regardless of their top speed, a bicycle will get you there almost as quickly.

WHO NEEDS PETROL?

After test-driving the Vectrix electric scooter, Johnny Borrell the Razorlight singer and ambassador for Friends of the Earth, said: 'You could certainly beat all the traffic in London on that thing.'

FLYING

According to 2005 government figures, aviation was responsible for just 6.3 per cent of the UK's carbon dioxide emissions. So why all the fuss about flying?

The reason is that this apparently small figure – seemingly a mere drop in the rising tide of carbon emissions – is not only more significant than it appears, but is also growing quickly and will soon take on gargantuan proportions. Other gases pumped out by aeroplanes flying at high altitudes at least double the impact of the carbon dioxide alone, so the impact of flying is really the equivalent of 13 per cent of the UK's total emissions. Even then it's argued that this doesn't paint a realistic picture of the emissions we're responsible for because the government's aviation statistics only count flights leaving the UK. If we were to include the return flights made by British citizens alone, but not flights made by foreign nationals, then aviation's share of carbon dioxide emissions in 2005 was closer to 18 per cent.

Planes have opened up the planet more than any other form of transport, and flying has become more and more affordable – partly thanks to enormous tax breaks enjoyed by the aviation industry (notably not paying VAT on fuel). With weekends virtually anywhere across Europe as cheap (if not cheaper) than a trip to the Lake District, previously out-of-reach destinations are now on our doorsteps.

This is great news for those of us looking for new and exciting experiences in far-flung places, but not so wonderful for the environment. Short-haul flights (less than 500 km) produce

triple whammy

At altitude the exhaust fumes from an aeroplane are 2.7 times more harmful than carbon dioxide on its own. Nitrous oxide and water vapour form ice crystals high in the atmosphere that trap the Earth's heat further. These all combine with carbon dioxide to create a nasty cocktail of greenhouse gases. So while carbon dioxide from flying officially accounts for just 6.3 per cent of the UK total, the impact is, in reality, more than double that amount.

FLYING GAME:

If plane trips are tempting, your carbon footprint will be growing. Deciding to fly less often will be a big step towards a greener life.

three times more carbon dioxide per person than the equivalent distance by rail and one long-haul return flight to somewhere like South Africa or Florida can produce more carbon dioxide per passenger than the average UK motorist creates in a year.

Because aeroplanes burn fuel at such a colossal rate – and at such high altitude – flying is an enormous factor in your own carbon footprint. No matter how carbon-careful you are from day to day, a few flights each year can very easily undo all that hard work. If you fly a lot and you really want to slash your carbon footprint, you're going to have to think about flying less.

fact

One return trip to New York accounts for 1.2 tonnes of carbon dioxide (3.3 tonnes of carbon dioxide with emissions multiplier) or 30 per cent of the average yearly emissions for someone living in the UK.

case study

'try living on a flight path, and then tell me flying is OK'

'In 2005 I spent five months in and around Siena. My trip started in February and ended at the beginning of August. I traveled out by Eurostar to Florence and returned the same way with my husband who had flown out to join me for a week's holiday. I chose to go by train because I campaign against expansion at Stansted Airport, and I do not like all the hanging around, or cramped conditions in flying. I was also not in a hurry, which might seem like a luxury to some, but if you make time and want to see the countryside from the window and have more space to relax, the train is the only way to go.

'I was traveling to improve my Italian so was glad of the time to adjust and to read. It is definitely more comfortable than flying budget airline to Europe. It might be a bit more expensive, but look at the damage aviation does to the environment, consider what mass travel does to some beautiful destinations, and actually live under a budget airlines flight path, and hopefully you'll soon change your mind.'

Suzanne Walker, Friends of the Earth 'green stories' contributor

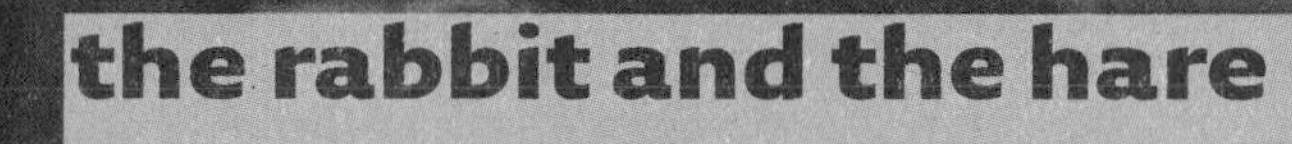

Six reasons why travelling by train is better all round.

1. It's greener. A return trip from London to Newquay produces 135 kg carbon dioxide if you fly, but just 72 kg if you go by train. Saving: 63 kg carbon dioxide.
2. The train takes you where you want to be. Stations are nearly always in the centre of town; airports are usually on the outskirts, making another journey essential to get to your destination (and adding more carbon).
3. You can watch the world go by, not just clouds.
4. You can stretch your legs, sit opposite someone and have a proper chat, surf the internet and use your phone.
5. You can take as much luggage as you like, including your bike, often at no extra cost.
6. You don't have to check in hours ahead.

how much carbon dioxide do my flights create?

Sarah Oliver, 26, a marketing manager living in Kent, took three flights last year all from London airports. Here's what her carbon dioxide emissions looked like. (In the table below a 2.7 multiplier has been included to take account of the greater impact of aviation emissions.)

Flights	Emissions (kg CO_2)	With multiplier (x2.7) (kg CO_2)
Weekend break to Poland	434	1,172
Business trip to Dublin	168	454
Two-week holiday to Cape Town, South Africa	2,173	5,867
TOTAL	**2,775 kg (2.77 tonnes)**	**7,493 kg (7.49 tonnes)**

Sarah's flights (even before using the emissions multiplier) account for almost a third of the average person's annual carbon emissions, or the equivalent of all the carbon emissions her home produces each year.

keeping our feet firmly on the ground – how the government should help

A £49 flight to Prague: tempting, isn't it?

Huge air transport growth isn't inevitable or necessary – the government can manage demand and develop the alternatives. Airlines and airports should pay for the problems they cause. The UK Government needs to rethink its destructive aviation policy and do the following:

- Say no to airport expansion. Why spend billions on new airports and runways when the real answer is to be cutting the number of flights being made, not facilitating their growth.

- Police the aviation industry. Voluntary solutions suggested by the airlines are inadequate. The government needs to oblige these companies to combat and reduce emissions. Let the industry grow unchecked and we'll see huge increases in emissions at a time when we should be making big cuts.

- Make the aviation industry pay fuel tax. The aviation industry manages to skip around £9 billion worth of fuel tax. Do you think it's right that one of the most polluting industries gets subsidised fuel?

- Increase Air Passenger Duty every year. Increasing Air Passenger Duty annually would help to manage demand in the interim before a strong enough emissions trading scheme can be introduced.

offsetting your emissions

The concept is simple. You work out how much carbon dioxide you generate at home or by travelling and then pay for this damage by buying so-called carbon credits. The money is meant to go to things like tree planting, environmental projects in the developing world and investing in renewable technology.

Offsetting has been promoted as a pain-free way of tackling climate change. Dozens of schemes now claim that carbon-conscious individuals and businesses can compensate for their emissions by donating in this way. And the rise of offsetting may have had some benefits – for example, helping raise awareness about climate change. Some of the companies involved have offered simple, at-a-glance carbon calculators that clearly illustrate the impact travel has on the planet; and 'gold standard' offset schemes target projects around the world, supporting poorer communities to access safe, renewable energy and adapt to the impacts of climate change.

Sounds great, doesn't it? Buy yourself out of the problem and carry on polluting the atmosphere. Unfortunately, it's not that easy. Offsetting schemes may look like a 'get out of jail free card' but they're no real substitute for reducing the amount of emissions we generate. Environmental organisations including Friends of the Earth have raised some serious doubts, saying schemes are simply being used to ward off legislation and are delaying the urgent action needed to cut emissions and develop low-carbon solutions. The fact is we just can't buy our way out of the problem – there is no solution to climate change other than to emit less pollution.

If you do want to help with these kinds of projects, support well-known charities such as Practical Action or Christian Aid; and if you do want to use an offsetting scheme, go for one that promotes energy efficiency or renewable energy with the gold standard accreditation. Three of the best companies are www.myclimate.org, www.atmosfair.de and www.climatefriendly.org.

fact

www.noflytravel.com is a travel site specialising in getting you to where you want to be without flying. It covers most of the world, showing train, car and boat options, and provides links to make planning and booking your trip as simple as possible.

"Carbon offsetting schemes are being used as a smokescreen to avoid real measures to tackle climate change. We urgently need to cut our emissions, but offsetting schemes encourage individuals, businesses and governments to avoid action and carry on polluting. There is still time to act, but we cannot afford to be distracted by measures that at best have only a small role to play in providing the solutions to global warming.

Tony Juniper, Friends of the Earth director

case study

the train to Spain is greener than the plane

Concerned about her carbon footprint but not wanting to miss out on a friend's thirtieth birthday celebrations (and a music festival) in Barcelona, fundraiser Louise Krzan, 29, from East London, ditched the plane and persuaded two friends to take the train with her instead.

'There's no doubt the journey took longer, but it was nicely broken up - mainly with pit-stops in restaurants and bars and the odd visit to world-famous tourist attractions. We caught the Eurostar from Waterloo mid-afternoon (check-in time was nine minutes) and three hours later we were making our way across Paris to Austerlitz Station, where we had a quick bite to eat before boarding the Spanish-run TranHotel. The sleeping rooms were small but

we'd been tipped off by a friend, so dumped the bags and headed to the bar carriage, next to the rather posh restaurant car, where we secured the three best seats in the house. A sandwich, several beers and numerous games of cards later we hit the sack, our beds had been turned down and there was nothing left to do but sleep. I won't pretend it was the soundest night's rest I've ever had, but at 8:30 a.m. on the dot we pulled into Barcelona's França station and strolled to our hotel that was just a few streets away – no queues, no shuttle buses, no stress and a lot less carbon.'

Cost:	£59 Eurostar return to Paris, £108 Paris-Barcelona (standard class)
Train emissions:	76 kg carbon dioxide
Aeroplane emissions:	205 kg carbon dioxide

tunnel vision: making the most of the train

Forget flying, the train makes eco-friendly foreign holidays easy and affordable. Here are some great destinations.

Alpine adventure:
Take the Eurostar to Paris and then the TGV to Geneva. From there, relax on board the Glacier Express that takes you on a leisurely journey over the Alps to glamorous St Moritz. An eight-day unlimited Swiss Rail Pass costs from just £164. (www.raileurope.com)

Enjoy the Craic:
It's surprisingly easy, wonderfully scenic and dirt cheap to get to Dublin by train and ferry from all over the UK. It costs from just £20 each way, and that's a standard fare. You can also pick up bargain deals including Manchester to Dublin for just £20 each way from www.sailrail.co.uk. While it does take a while (ten hours), there's some beautiful countryside to enjoy – sit back and relax.

Say hello to Saint Peter:
A luxury train journey to Rome makes the city all the more romantic. Catch the lunchtime Eurostar and change in Paris for the Palatino sleeper to Rome. On board you can relax in the restaurant (not a packaged sandwich in sight) before retiring to your sleeper car. When you wake you can enjoy the Italian countryside before arriving fully refreshed in the heart of Rome at 9.40 a.m. Eurostar costs from £59 return and a two-bed sleeper to Rome costs from £104 each.

CYCLING

It's quicker than walking and quicker than driving in many towns and cities. Kids love it, it's cost-effective, good for your health and completely carbon dioxide free. Many people think of cycling as their main mode of transport and only use alternatives where absolutely necessary. Chances are, unless you live miles from your nearest neighbour, you'll save time, the school run could actually be fun and you'll be fitter, have more get-up-and-go and more money.

For every five-mile trip you cycle rather than drive you're saving 1 kg (based on a 200 g/km car) of carbon dioxide. Cycle rather than drive every day to work (and back) and you'd save 466 kg a year. If you persuade your next-door neighbour to do the same, that's 1 tonne of carbon dioxide saved between you – not to mention fewer cars on the roads, less exhaust fumes and fewer accidents.

case study

'there's just so much to explore in this fair isle of ours'

'This year (2007) I took my kids out of school and set off to show them Britain. I had three months to travel around but it soon became obvious that there's so much to do on holiday here that I decided to just explore the north – a lucky decision given that much of southern England was flooded in 2007. Up here the countryside is so varied and many of the cities are places arty people just must visit – such as Newcastle and Glasgow. But I also love the lesser-known places like farm shops and locally-owned delis and pubs. You can try fantastic regional food – nettle cheeses in Northumberland, raspberry jam in the Borders, fish along both coasts and locally-brewed beers all over the place (the Lake District town of Cockermouth is the star). There are footpaths and picnic sites to explore if you get tired of admiring the historic sites opened up by English Heritage, the National Trust and private families. We've seen castles, ruins, gardens, sculpture parks and tried out the walk along Hadrian's Wall and the sandy beaches of Northumberland. We've swum in Stonehaven's heated sea water lido, fed seals at Eyemouth harbour and climbed our first Lake District mountain, Skiddaw. But we've also enjoyed having the time to dawdle; to walk the long way through the park just to get the best ice cream, watch the boats or listen to the oyster catchers, and admire the evening sun illuminating the fields of Haltwhistle on the cross-country Newcastle to Carlisle train. Jetting off makes no sense when the pleasure of staying at home – so long as you treat yourself to a holiday mood and good quality waterproof clothes – can be so great.'

Nicola Baird, 43, London, co-author Save Cash & Save the Planet (Collins)

case study

"What's not to like?

Commodity manager Gareth Ewens, 31, cycles from Walthamstow in East London to his office in Canary Wharf, a journey of about eight miles each way. 'It took a little getting used to at first, but I use my work gym to shower and I feel far more relaxed and awake when I start work,' he says. 'The tube used to wind me up – I just hated sitting still in tunnels every day, but now my journey takes about half an hour less, saving me £30 a week and keeps me fit. What's not to like?'

Gareth's carbon savings: 16 miles (25 km) per day

...**Drive** (based on a 200 g/km vehicle): It will take around 36 minutes (assuming no traffic in central London) and produce 5 kg carbon dioxide per day, or 1.16 tonnes per year

....**Get London Underground** (assuming the distance is the same as by road): It will take around 106 minutes and produce 1.4 kg carbon dioxide per day or 325 kg carbon dioxide per year.

...**Bike:** It will take just 45 minutes, cost nothing to ride and produce zero emissions.

cycle to work

As part of the government's plan to get more people to cycle more, the Cycle to Work scheme allows an employer to loan bicycle and safety equipment to employees at up to 50 per cent off. Your employer buys the equipment and loans it to you as a tax-free benefit – basically, the cost of the bike, minus VAT or National Insurance, is then taken from your salary over 36 months. After this you have the option of buying the bike outright at a fair market price.

The chances are that if you have a folding bike you won't have any trouble getting on most forms of public transport, but for everyone else it's worth checking before you travel as many main routes still won't accept bikes at peak times. Things are getting better, slowly. A great source of information is A to B

FAB GEAR:

The bicycle – it's regularly voted the best invention of the past 150 years, and some 2.5 million are sold each year in the UK. There's never been a better time to get in the saddle.

www.atob.org.uk,whose transport list is updated regularly and even includes ferry companies and coach services. Alternatively, call National Rail Enquiries on 08457 484950, www.nationalrail.co.uk.

buying a bike

If you're going to be using your bike every day, it's worth spending time – and some money – getting something that's right for your needs.

Find a shop with a good reputation. Ask friends who cycle. You want somewhere with helpful, knowledgeable staff who have time to help you.

Think about what cycling you'll be doing – road, mountain biking, commuting, pootling around town.

Spend as much as you can afford. The sky's the limit with how much you want to pay, and serious cyclists will always recommend spending as much as you can afford. You can pick up a very basic new model for £80–£100, but it may not be as enjoyable to ride or as long-lasting; you might be better off looking for a better-quality second-hand design. For a new hybrid expect to pay £200+, a good mountain bike will cost £300+ and folding commuter bikes start from £300-400.

fact

Have your cake and eat it. A half-hour commute by bike will burn eight calories a minute, or 11 kg fat a year.

lifecycle

So we've established that bikes are healthier, cleaner and (if you live in a city) often quicker than cars. If you're a bike owner, it's also worth patting yourself on the back because of the much smaller quantity of energy it takes to manufacture a bike, its low maintenance costs, and (with a little care) its longevity. For every car on the road 10-20 per cent (depending on the model) of the carbon it produces in its lifetime is generated during the manufacturing process. And when you think that a car can easily travel more than 100,000 miles in its lifetime its fossil fuel burn is enormous.

types of bike

Mountain bikes: Designed for off-road riding, but they're very popular for everyday use, are easy to ride and have plenty of gears. The big knobbly tyres make pedalling on the roads a bit harder.
Hybrids: A cross between racers and mountain bikes, these are light, have smooth tyres and usually an upright riding position that's particularly suited to urban cycling – a good choice for everyday use.
Folding bikes: Useful if you commute to work by bus or train but want to cycle some of the way. The latest designs fold up extremely small and are surprisingly easy to use, but the smaller wheels and limited range of gears mean they are only suitable for shorter trips. Sales have increased enormously over recent years with bike shop Evans seeing a rise in demand of almost 75 per cent over just one year; the most popular folding bike, Brompton, has a waiting list of at least 12 weeks for a basic model.

second-hand cycles

You don't have to splash the cash to get on your bike. Here are some good alternatives.
Visit a police auction: Unclaimed stolen bikes are sold off at these auctions for next to nothing. The bikes are sold as-seen so you won't get a guarantee or a test ride, but you might well pick up a bargain, or some useful parts at least if you're going down the DIY route. Ask at your local station or check out www.bumblebeeauctions.co.uk for details. You will need to collect, so it's worth bidding locally.
Try online: Property sharing website www.freecycle.org often has old bikes up for grabs for free, but for the largest selection of discounted bikes nothing touches eBay. When we checked it had no fewer than 3,374 bikes for sale, not to mention thousands of parts and accessories.
Check the shed: Bike shops can work wonders even on the oldest, creakiest frames, so if you know anyone who's got anything they don't use tucked away in the garage or shed, ask nicely and get it to your local bike shop for a full service

on yer bike –
how investment in cycling from the government would slash congestion, health costs and carbon

Recent analysis by cycle charity Sustrans has revealed that every £1 spent on a cycle route generates benefits worth £20; the average return of rail and road schemes is typically £3 for every £1 spent.

Cycling England estimates that a 20 per cent increase in bicycle journeys would lower healthcare costs and reduce congestion – a cyclist costs the NHS around £30 less each year.

Cycling England has estimated that a £70 million annual investment in cycling initiatives by the government could cut up to 54 million car journeys a year by 2012 and reduce carbon dioxide emissions by 35,000 tonnes. Since 2000 the National Cycle Network has carried almost 1 billion journeys representing a potential carbon saving of more than 1.3 million tonnes.

Encouraging 20 per cent more cyclists on to Britain's roads could save the taxpayer more than £520 million – and help fight climate change.

the essential cycling kit – be safe, be seen

Helmet: Be sure it's certified to EN 1078 standard.
Lights: Between sunset and sunrise you'll need one white lamp on the front and one red lamp on the back. Energy efficient LED bulbs are ideal.
Locks: Use two different types of lock (one chain and a D-lock) – chances are a professional thief will have only bought equipment to break one type.
Clothes: Always remember a reflective jacket. If you don't want to spend more than a couple of pounds, try your local builders merchants or hire shop (HSS sell a selection).
Bell: All new bikes should come with a bell as standard. If you don't have one, go and buy one – you'll thank us.

cycling with children

Whatever your family members' ages there's equipment out there designed to help you all get around.

Baby Seats (6 months to 5 years, 1 passenger):
Fixed to either the front or back of an adult bike, these will allow you to take your child out and about, even on longer journeys. Check that the weight capacity of the seat is suitable for your child (too heavy will not only put the seat under stress, but make your job as chauffeur much harder), that their feet are kept well clear of the wheels and that there's a good harness with at least two straps over the shoulders, one between the legs.
Trailerbikes (4 to 9 years, 1 or 2 passengers):
Also known as Tag-A-Longs, these small or often half bikes offer you the opportunity to share your cycling experiences with your child. Attached to your bike with a tow-bar, they let your children practise being on a bike, and if you choose one with pedals they can help you out with the work, or freewheel and

enjoy the ride. A good choice when you have one (or more) children old enough to ride their own bike, and a younger child who doesn't want to miss out. Always take a trailerbike for a test ride before you decide. It has to be manageable for you, and comfortable for the kids. You can also find devices (Gator Arms or Tow-me's) that attach a kid's bike to an adult bike's frame, which makes longer rides much more comfortable – for the little one at any rate. For more information visit www.whycycle.co.uk/children-tagalong.htm.

Trailers (birth to when they grow out of it,1 or 2 passengers):

Turn your bike into a carbon dioxide-free MPV. With room for two children, and offering complete shelter from the weather, trailers are a great alternative to the traditional school run and when you've dropped the kids off you've got plenty of space for

CHAIN GANG:

With health and fitness high on the agenda and provision of cycle routes continuing to expand peddling is enjoying a boom.

the grocery shopping. If you'd feel more comfortable having the kids up front where you can see them, specially designed tricycles are available (not too dissimilar to a rickshaw). Costing from £1,000, these are not cheap, but compared to the running cost of a car they're great value for money.
Price: Child seat £25-£100; Trailerbikes (Tow-A-Long) £150-£300 ; Trailers £160-£500; Tricycles £1,000-£2,000.

Useful contacts:
www.orbit-cycles.co.uk sell the Bluebird range of trailers including the flexible 5-in-1 design which cost from £230.
www.nihola.info Front seat trailer designs from Denmark. Costing from £1,200-£2,000 they're not cheap (but remember you could save 50 per cent with the Cycle to Work scheme)
www.bikefix.co.uk Huge selection of trailers, bike seats, tow bars.
www.bikesandtrailers.com Great selection of bikes, tow-a-longs and trailers
www.amba-marketing.com Vast selection of innovative European designs including front fitted child seats.
www.tricycleweb.com Innovative three-wheelers for transporting people and heavy loads quickly and safely. Prices from €499.
www.triobike.com. Kiddie carriers for the style conscious family with the added benefit of transforming into a two wheeler when needed.

top tip

Sustrans the cycling promotion charity has set up the National Cycle Network which now boasts more than 10,000 miles of paths. At present 75 per cent of the population live within two miles of a route. Visit www.sustrans.org.uk for details. If you're in London Transport for London print cycle maps covering the whole city.

case study

'no more gym for me'

'For years I would blindly get on the bus or the tube like a sheep and end up at work without really thinking about it, and the thought of cycling didn't even cross my mind. Then we moved to within just a few miles of work and I thought it was silly not to give it a go. As it turned out cycling takes about half the time public transport did – about twenty minutes instead of forty. My partner Simon and I now cycle every day and it's just part of the routine. OK, if we're off out after work we might leave the bikes at home, but I'd say we cycle 80 per cent of the time. I've got no fancy bike gear, just a cheap bike with a basket, but it gets me to work just fine. Yes it's pretty horrible when it rains but we've got changing rooms at the office and a shower if I really need one.'

Nicole Story, publisher, Hackney, East London

cycling to school

Between 1994 and 2004 there was a 31 per cent increase in the number of people using a car for the school run, but if you've ever had to do one you'll know just how useless your car is at 8:45 a.m. and 3:10 p.m. when every other parent descends on the same few streets.

Cut out the stress, hassle and needless emissions by getting your kids to cycle to school, and increase their (and your) fitness, alertness and general wellbeing every morning. Cycling charity Sustrans has established the Safe Routes for Schools programmes to help parents, teachers and pupils to change the way they get to school. Ranging from cycle convoys and walking groups to the funding of more cycle paths, bike sheds and cycle safety training, there is bound to be something suitable for your child's school. Visit www.saferoutestoschools.org.uk or call 0117 9150100 for an information pack for your school.

Also run by Sustrans is the Bike It campaign, funded by the bicycle industry and backed by Cycling England with the support of the Department for Transport. Its main role is to tackle all the obstacles that get in the way of more children cycling safely. It ensures that there is a safe route to school, organises cycle training to the National Standard, and even arranges secure cycle storage. If you'd like to get involved and see how a Bike It Champion could help increase cycling rates at your child's school visit www.sustrans.org.uk/askforbikeit.

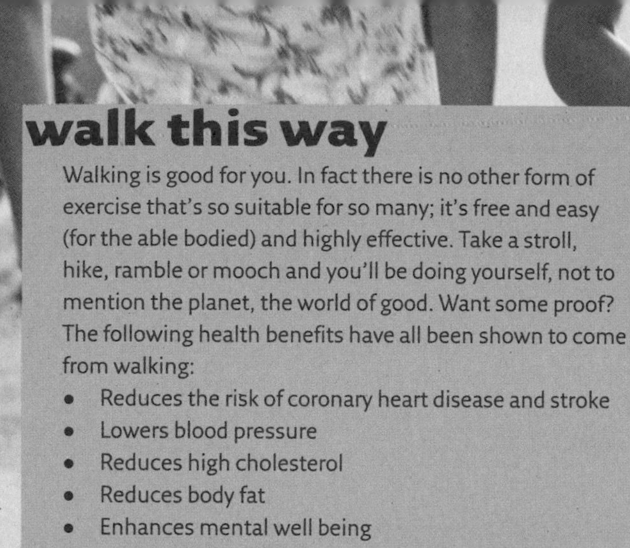

walk this way

Walking is good for you. In fact there is no other form of exercise that's so suitable for so many; it's free and easy (for the able bodied) and highly effective. Take a stroll, hike, ramble or mooch and you'll be doing yourself, not to mention the planet, the world of good. Want some proof? The following health benefits have all been shown to come from walking:

- Reduces the risk of coronary heart disease and stroke
- Lowers blood pressure
- Reduces high cholesterol
- Reduces body fat
- Enhances mental well being
- Increases bone density, hence helping to prevent osteoporosis
- Reduces the risk of cancer of the colon
- Reduces the risk of non insulin dependant diabetes
- Helps to control body weight
- Helps osteoarthritis
- Helps flexibility and co-ordination hence reducing the risk of falls.

With about a quarter of all car journeys less than two miles long, chances are that once you've got stuck in traffic, found a parking space, fumbled for change and paid-and-displayed you'd have been able to walk there without polluting the atmosphere or paying for the privilege. And for the calorie counters amongst us walking one mile can burn at least 100 kcal. Walking two miles a day, three times a week, can help reduce weight by a pound every three weeks. Not a bad return for your effort.

PUBLIC TRANSPORT

The maths is easy. The more people in the car, bus, train or plane, the more energy-efficient that journey becomes. Leave the car at home and use public transport and you'll greatly reduce your personal carbon dioxide output. A journey by train or bus typically creates less than half the amount of carbon dioxide as the same trip in a car, not to mention taking the pressure off you as driver or navigator. It reduces the stress on our overcrowded roads, too, cutting accidents, noise and air pollution. And who doesn't crave the chance to catch up on a good book or just watch the world go by?

train travel

If you drive from Manchester to London, say – a journey of 200 miles – you'll generate 52.9 kg carbon dioxide and it will take 3 hours 52 minutes. By train you'll be responsible for just 12.9 kg carbon dioxide, it will take you 2 hours 24 minutes and you won't need to pay the congestion charge (£8 as of 27 September 2007), find a parking space, or negotiate the M25. Ironically the overcrowded commuter train is even more carbon-friendly – a small but important comfort when you're squished into a stranger's armpit.

top tip

If you've ever used the line 'I'd get public transport but I can't' as an excuse to jump in the car, check out www.Transportdirect.info. It's a free journey planner for routes all over the country – roads included. It might not show you the quickest route, but it will be the greenest available.

trams

They're not a new idea, but inner city trams are still quick, convenient and extremely low carbon. Nottingham, Manchester, Birmingham, Sheffield, Croydon, and Blackpool all have reliable and effective networks that produce around 65g carbon dioxide per km – better than traditional rail and leagues ahead of the car.

coaching routine

For great value low-carbon travel the national coach network is hard to beat. The average coach trip can produce just 28 g per passenger per kilometre, which is significantly better than both rail and the car and the prices for tickets are substantially cheaper – look at www.megabus.com for low-cost fares. National Express has over 1,000 destinations across the UK and its buses cover hundreds of thousands of miles every year. Unfortunately you're still at the whim of traffic jams, delays, diversions and long journey times but you'll save substantially compared to running a car.

welcome to the cheap seats - getting the best deal on train tickets

- Book early. Network Rail has to set the national timetable 12 weeks in advance and this is often (but not always) when tickets are released. Try to book 12 weeks ahead to get the really cheap fares.
- Two singles might be cheaper. Doesn't seem right, does it? But buying separate single journeys frequently costs less than a standard return so it's always worth checking before you buy.
- Call if you're not sure. If you can't find an advertised fare online or if prices seem too steep, call up the rail company and ask. The operators know their stuff and might be able to find you a cheaper route.
- Beat peak time trains. If your journey includes several changes at major stations, it's worth checking the prices of individual tickets to each station as there may be special fares available for some of your trip. Also, if you're beginning your journey at a peak time but know you're going to change trains after peak hours, it may work out cheaper if you buy two tickets, one peak, and one off-peak.
- Always look out for railcard deals. National railcards typically give you a third off all ticket prices so they're well worth investing in. If you're paying more than £60 for your ticket, buying the railcard effectively pays for itself. Also look out for student, young person and senior citizen cards, family rail cards, where kids and adults get substantial discounts when they travel together, and regional railcards, which allow cut price fares from area to area.

remote access

Alas, we're not all blessed with a regular public transport system, fancy trams or local rail networks to keep us out of our cars, but there are community schemes cropping up to help some of us who live in areas where the transport providers can't justify running a 'public' service.

Devon: There are currently six local, and non-profit, mini-bus routes for areas surrounding towns like Totnes and Tavistock that provide a regular service to areas only usually accessible by car. There is also a scooter loan service available for people who need to travel to find training or work.

Cornwall: CORLINK is a network of scheduled and flexible vehicles in the Wadebridge and Bodmin area designed to make rural travel easier. Locals can book a journey in advance and can be picked up and dropped off within designated zones, all for the price of a bus rather than taxi.

Staffordshire: Grant aid is available as part of a scheme to encourage people to develop their own transport initiatives to reduce rural isolation and help provide affordable transport. The Rural Parish Local Transport Grant Aid Scheme is available to groups helping to extend the local public transport services especially for those who don't have (or don't want) to drive.

chapter 9

scaling up your impact

At work, at school in the neighbourhood –
find power you never knew you had.

beyond your front door

If you really want to make a difference, you'll need to look further than your own front door. Schools, workplaces, sports clubs, scout huts, pubs, etc., are all voracious consumers of energy, but they're spaces we need and want. You can help them make a difference – from simply persuading friends to turn their lights off to campaigning to getting the council to cut the carbon footprint of its own buildings. This chapter looks at how to get started.

WORKING TOWARDS A LOW CARBON LIFE

If you've followed the practical steps in Chapter 7 you're hopefully well on your way to creating a low-emissions home that costs you little to run and won't cost the earth.

It's a shame then that at 8 o'clock every morning the vast majority of us leave home to spend our days in draughty inefficient workplaces that contribute millions of tonnes of carbon dioxide to the country's footprint.

The irony is that the improvements you've made at home can just as effectively be applied to the workplace. There may be some red tape and the odd sceptic out there, but you'll be amazed at how quickly they'll change their tune when they realise how much money they could save. Envirowise, the government's environmental business advice service, estimates that there could be potential savings of £3 billion just through making environmental improvements.

de-carbonising your workplace

Here are eight things you can do today:

1 Switch off plugs. Shut down your computer, switch off your monitor and if you're last out turn off the printer, photocopier, turn off the radios, microwaves and phone chargers. Some well-made devices draw just milliwatts of current on standby, others will suck up 30 watts from the mains, or 113 kg carbon dioxide per year without it doing anything useful. Leaving the photocopier on overnight uses enough energy to print more than 1,500 A4 pieces of paper.

2 Turn out the lights. How often do you walk by a building at night and see all the lights on? It takes just ten fluorescent fittings left on for eight hours a day for a year to create 1 tonne of carbon dioxide. What about the other 16 hours when no one's in the office? That's a further 2 tonnes of carbon dioxide if you don't turn them off.

3 Get stickering. Print out a few sheets of 'Switch me off' stickers and put them around the office on lights, monitors and printer/copiers. They're free from the Carbon Trust and you'll be amazed how (as much as the office sceptics thinks it pains them to do so) quickly it gets people thinking, and switching things off.

4 Use the energy-saving mode. You should be able to turn the brightness of your screen down, or to turn it off if you haven't used it for a few minutes. For instructions on how to do this on PCs or Macs visit www.nrel.gov/sustainable_nrel /energy_saving.html.

WHITE COLLAR CRIME:

According to the Carbon Trust, offices account for 6.8 million tonnes of carbon dioxide per year and a 20 per cent saving in the amount of energy used could save UK businesses £157 million.

5 Get the bus. Make the most of public transport. Ditch the car and take the bus and you'll typically save 688 kg carbon dioxide per year. Take the train and save 607 kg carbon dioxide, or buy a bike, get fit and save 854 kg.

6 Print less. Do you really need to print all your documents – for internal meetings, for example? And if so, can print on both sides of a piece of paper? This will effectively halve your paper costs. Check with the printer manufacturer to see if your machine is compatible. If you can, set it to print double-sided as standard.

7 Fly less. Do you really need to travel to a business meeting? Think about using the latest conference-calling phones and web cams – they're much easier to use and sound better than they used to. Just one 400-mile round trip by plane adds 1.38 tonnes of carbon dioxide to your personal footprint so only fly when absolutely necessary, and choose the train where possible.

8 Buy a mug. If you're still using paper, plastic or polystyrene cups for tea and coffee, go and treat yourself to a nice china mug. The tea will taste better, stay hotter and you won't be contributing to landfill every time you take a break.

top tip

If your boss only thinks in pence and pounds try reminding them that: for many businesses a 20 per cent cut in energy costs represents the same benefit as a 5 per cent increase in sales.

turning the boss green

While many companies are now realising that low-impact and sustainable is the way to go, not all have been quick to act. If you want to make a difference to the way your office or building is run but don't think you can do anything alone, please think again. There's plenty we can all do to help persuade our bosses that cutting emissions isn't just good common sense, it makes perfect financial sense too. Here's how to get started.

Get a plan of action.

The Carbon Trust offers a free online service called 'Build Your Action Plan' listing a range of effective energy-saving methods for your business. Simply pick your business sector (from hospitality to engineering and everything in between) and the size of your energy bill and the calculator shows a range of proven energy-saving options, how much they'll cost and what sort of savings you could make. It's a quick, free and a positive first step. For more visit www.carbontrust.co.uk/atp.

The bottom line.

What will climate change cost your company? Increasing oil, gas and electricity prices, government restrictions on emissions and the Climate Change Levy (a tax on non-domestic electricity consumers) will all add up and dent profits. Acting now and managing the risks will save later on. Remind management that low energy = low cost.

Give examples.

Egos and one-upmanship are just part of most businesses, so play to them. Tell your boss about some success stories (see the case studies below) and highlight not just the energy saving but also the positive PR they've got out of it. See what your competitors are up to: nobody likes being left behind. To find out what 500 of the world's top businesses are doing about climate change visit the Carbon Disclosure Project website www.cdproject.net where you can find out the amount of carbon dioxide emissions they produce, how much they're reducing them by and what changes they're implementing (if any); and if your company's involved you can see how you compare.

Keep it positive.

It's very easy to slip into being negative when talking about climate change, but you won't get very far by laying on a guilt trip. Much better to avoid the doom and gloom language associated with global warming and talk about the benefits of going carbon-neutral instead.

Get a man in.

If your company has an energy bill that exceeds £50,000 the Carbon Trust will conduct a free on-site survey of its energy use (and abuse) and can offer a tailored package of solutions to make your workplace more efficient. There are even interest-free loans available to help kick-start a low carbon programme.

calling in the professionals

The following organisations offer practical advice, loans and information on where energy and environmental savings can be made at your workplace.

The Carbon Trust:
This not-for-profit government-funded company helps businesses and the public sector cut carbon emissions and supports the development of low-carbon technologies. It offers free phone and on-site consultations, enormous amounts of information and interest-free loans for carbon-saving investments. For more information visit www.carbontrust.co.uk.

Envirowise:
A government-backed scheme offering UK businesses free, independent, confidential advice on practical ways to boost profits, minimise waste and reduce their environmental impact. They run conferences and workshops and can tailor carbon-cutting activities to suit your business. For more information visit www.envirowise.gov.uk. Free helpline on 0800 585794.

Association of Town Centre Management:
This company wants you to become a Business Climate Champion and pledge to save at least quarter of your energy. Aimed at small and medium-sized companies it can show the best way to make positive changes. If your business meets its targets its name will be added to an official Roll of Honour, and be entitled to use a logo that demonstrates to staff and clients they're making a difference against climate change. For more information visit www.businessclimatechampions.org

Global Action Plan:
Environmental charity with a range of schemes to help businesses cut emissions and waste. One GAP scheme, EcoTeams@work helps businesses set up a social network among employees who want to reduce their impact on the environment and save money. Groups of six to eight are encouraged to meet once a month to decide on the environmental actions that they will take at home, and share experiences of what they've already done. Results: lower bills and less waste at home while the business gets benefits such as raised environmental awareness and greater interaction with staff. For more information visit www.globalactionplan.org.uk

case study
Pfizer

Based in Sandwich and Walton Oaks in Surrey, Pfizer has a nationally-recognised green transport scheme to help cut its carbon footprint. 'The scheme actively encourages all our employees to make use of the bus services we provide instead of using a car,' explains Johannes Waltz, Pfizer's stakeholder strategy manager. 'We provide discounted rail tickets and a shuttle bus to take employees to key drop-off points throughout the day – and we also guarantee a lift back home in an emergency at any time of day.'

case study

'having an ethical agenda has really helped to engage our employees'

Kim Stoddart is the director of Hove-based media and PR agency the Green Rocket Group. Her company helps small ethical and green businesses get their message out to a wider audience.

'Green Rocket is a social enterprise business developed to help give back to the community; and as a result 75 per cent of our profit goes back into local projects. Everyone who works here works half a day a month on a local sustainable project,' explains Kim. 'I became increasingly frustrated with the blatant greenwashing that was going on in big business and felt that the smaller companies who had really strong ethical and environmental credentials were not getting a fair look in.

'We're only a team of ten but I encourage everyone to look at the way we work, and if they can see possible improvements not to be shy about suggesting them. In one meeting we were discussing how we can save water and someone suggested we invest in a water-saving kettle. This sounded like a great idea until one member of the team piped up and simply drew one-cup, two-cup, three-cup lines in felt tip on the kettle we already had. It's only a small thing, but it has not only saved us £50, many kilograms of carbon dioxide and gallons of water but showed the team was really caring about where they worked and what they use.'

Kim and the Green Rocket team will only use companies with good green credentials. These include Good Energy, The Co-op bank and Green Your Office. They recycle office furniture, computers and mobiles and everything else that can be recycled is, including paper, glass, cardboard and plastic.

To see how they're getting on visit www.greenrocketgroup.com.

case study

ExCeL

The London International Exhibition Centre (ExCeL) in London's Docklands is the largest open exhibition space in Northern Europe; its electricity bill was a cool £1.1million. Peter Mooney, chief operating officer at ExCeL contacted the Carbon Trust, who carried out a site survey and suggested a range of carbon-saving initiatives (mostly no cost behavioural changes) that cut £350,000 off their electricity bill all for a total investment of £60,000. Not a bad return.

GOOD SHOW:
ExCel shaved a quarter off its electricity bill after advice from the Carbon Trust.

business transport

Transport is the only sector where emissions are expected to be higher in 2020 than they were in 1990. If your business or workplace relies on getting people or goods around, it's worth taking a look at how efficient your transport it. You'll be amazed how much carbon dioxide you can cut, not to mention the money saved.

The Energy Saving Trust (EST) offers a free fleet review to help lower running costs, reduce environmental impact and enhance your business's corporate social responsibility. If you have a fleet of 50 or more vehicles (20 in Scotland) under 3.5 tonnes you're eligible for a free review; if fewer than this there's a telephone advice line. The EST will help you cut your carbon footprint by assessing your transport needs, the mileage, and driving techniques and suggest practical ways to help cut carbon (and almost certainly costs).

To sign up for your survey visit www.energysavingtrust.org.uk/fleet.

did you know?

Heavy goods vehicles account for about 5 per cent of the UK's total carbon emissions.

be a green van man

Given that most vans have a typical lifespan of just five years, if your company has a fleet of vans some will be due for replacement soon and it's worth looking into the latest electric trucks that offer emission-free driving for towns and cities. Smith Electric is leading the pack with its range of electric vans and trucks that cut emissions by around 44 per cent compared to a similar sized diesel van. Each van saves around 5 tonnes of carbon dioxide per year, and while they are almost double the price to buy they're much cheaper to run than a typical diesel

	Smith Edison 3.5 tonne Long wheel base	Diesel 3.5 tonne Long wheel base
Vehicle Cost	£ 36,000	£ 12,300
Total Fuel Costs	£3,950	£ 23,400
Congestion Charge Based on 5 entries per week over 5 years based on current £8 charge	NIL	£ 10,400

model, are exempt from congestion charging, can park for free in many city centres and are almost silent – ideal for night-time deliveries (anyone remember the milk float?).

Both Sainsbury's and Royal Mail have started to use the Smith Electric vans. They may have a top speed of just 50 mph and only travel 150 miles on one charge but with the average rush-hour traffic outside London moving at 21 mph and just 12 mph in the London congestion zone, speed really isn't an issue.

For more details visit www.smithelectricvehicles.com.

company cars

There are around 3 million company cars in the UK and they're responsible for 16 million tonnes of carbon dioxide. The luxury saloon paid for by the boss is still seen by many in business as a benchmark of success, but to really cut your company's carbon footprint you should try to limit the choice of cars to low-emission models. We're not imagining that all the fat-cats will trade in their exec saloon for a G-Wiz electric two seater (although it would be nice) but if you can limit the choice to one of the 100 or so lower emission cars (such as the Toyota Prius and Honda Civic hybrids to the Citroën C1 the Vauxhall Corsa,

Renault Modus and Toyota Yaris) you could slash your company's carbon emissions, pay less Vehicle Excise Duty and pay less Company Car Tax.

driving down emissions

Do the things on this list and the Energy Saving Trust estimates that a fleet of 100 vehicles you could save £90,000 a year.

- Encourage the use of vehicles with low carbon dioxide emissions.
- Look after vehicles according to manufacturers' specifications – poorly maintained vehicles use more fuel and emit more carbon dioxide.
- Analyse the fuel consumption of your drivers and you'll be able to see who drives efficiently, and who would benefit from some training.
- Offer training to promote more efficient and economical driving.
- Provide journey planning advice – add a journey planner to your company's homepage and encourage employees to check their route, and where possible consider leaving the car at home (say on journeys under 3 miles).
- Invest in satellite navigation; it can help keep drivers out of traffic jams and ensure they get to their destination quickly.
- Encourage tele- or video conferencing to reduce the number of business trips needed.

top tip

To compare company cars, their efficiency and the tax your employees could save by choosing a low-emission vehicle, visit www.comcar.co.uk.

car clubs

If your workplace isn't as reliant on company cars but still needs access to vehicles from time to time, a work-based car club could be one solution. Creating a pool of vehicles for staff to use when public transport isn't a practical alternative will slash your transport costs (you only pay for what you use and have no direct maintenance or insurance costs, or huge unexpected garage bills), will make your employees think harder about their travel habits (especially if they have to plan ahead and book journeys instead of just jumping in and driving) and make a significant dent in the amount of carbon dioxide you're responsible for. Check out these established clubs that cost from £50 to join and then from £3-5 per hour to drive:

City Car Club – www.citycarclub.co.uk
Streetcar – www.streetcar.co.uk
Whizzgo – www.whizzgo.co.uk
Zipcar – www.zipcar.co.uk
For more details and an interactive map of all car clubs in the UK visit www.carplus.org.uk.

" Every Streetcar takes an average of 26 privately-owned cars off the roads of the UK which means we have already taken over 6,500 cars off the streets. What's more we've found that members choose to drive a car 68 per cent less after joining because they use other means of transport more.

Paul Johnson, Streetcar

working in London?

Consider a car that's exempt from the Congestion Charge to ensure lower emissions, reduced fuel costs and more money in the bank. For a list of the vehicles you can drive in London for free visit www.cclondon.com.

cycle to work

Create a fit, happy and low-carbon dioxide workforce by establishing a Cycle to Work scheme at your company and try and make a dent in the 341,000 tonnes of carbon dioxide per year that's created by the 78 per cent of commuters who still drive or are driven to work. A Cycle to Work scheme costs nothing to set up and can provide members of staff with tax-free bikes and safety equipment at typical savings of around 50 per cent. Your company buys the equipment up to the value of £1,000 and loans it to staff as a tax-free benefit – basically the cost of the bike, minus VAT or National Insurance is then taken from salary at source using PAYE over an agreed period, usually 12 or 18 months. If your boss doesn't think they have the time or resources to manage the scheme spell out the benefits:

Create a healthy workforce: On a day-to-day basis, better fitness means fewer sick days and better general health.

Staff will be happier: Cycling reduces stress and chances are your staff will turn up at work feeling bright and alive.

They'll be on time: There'll be no 'The train was delayed' or 'The B604 was at a standstill' excuses anymore.

You'll save money: Ten bikes can be securely locked in the space you need for one car, slashing parking costs.

You'll look great: Go on, be a nice boss, your workforce will be happier and more motivated to work for you.

Cut your footprint: For each employee who cycles there will be 854 kg less carbon dioxide warming the planet each year. Your company's reputation will be enhanced. Some 600,000 people already cycle to work, saving almost 5,000 tonnes of carbon dioxide each week.

work from home

Speak to your boss and, if your job allows, see if you can work from home one day a week. While this won't be possible with most manufacturing jobs, if you're desk-bound a broadband connection and phone might be all you need to help reduce your carbon footprint. It's not guaranteed, however, because if you walk to a café to pick up lunch at work, but jump in the car and drive to the supermarket when at home you're just cancelling out the savings staying home can make.

SUNRISE INDUSTRY:

If you work from home choose the brightest, sunniest room for your study. It'll cost less to heat and be comfortable throughout the day.

use green services

If you want a truly eco-friendly workplace, you'll need to think about using green services and suppliers. Here are some suggestions:

Environmental cleaning services:
Make sure your contract cleaning firm uses eco-friendly, non-toxic products. ICS Cleaning Services (www.ics-online.co.uk) covers London, Essex, Kent and the Home Counties and www.kingswoodecocleaning.com based in Croydon is one of only a handful of companies using organic chemical-free laundry cycles, energy-efficient equipment and non-toxic, water-based, hypoallergenic and biodegradable cleaners.

Stationery suppliers:
For a comprehensive list of companies specialising in recycled and environmentally friendly products visit www.recycledproducts.org.uk. For specific office supplies www.greenstat.co.uk (01225 480 556) sells a vast selection of goods, from recycled printer cartridges to eco cleaning products and tea bags.

Green taxis:
In London the two biggest players are the Green Tomato cab service (www.greentomatocars.com) which has a fleet of hybrids, and the black cab firm www.radiotaxis.co.uk, which operates 3,000 cars across the capital and which claims to be the world's first carbon-neutral taxi company thanks to investment in methane capture, wind-power generation and energy-efficient lighting.

Couriers:
Unless you're often transporting large items, make sure you use bike (cycle, not motor) couriers wherever possible. In a busy city centre they're usually quicker and you'll be saving the planet 4 kg carbon dioxide for every 10 miles. If you use ten couriers a day you could be saving more than 6.5 tonnes a year.

Printers:
Look for a company that uses Forest Stewardship Council (FSC) certified recycled paper and low VOC (Volatile Organic Compounds) ink from vegetable oil rather than mineral oil based inks. For help on finding the best printer for your companies needs contact paper.print.environment www.ppe.uk.net which is endorsed by Friends of the Earth, Greenpeace and WWF.

IT:
We can't avoid computers in the workplace, but we can make an informed decision about the companies we choose to buy from. In a Greenpeace survey (April 2007) the following companies were seen to be the least damaging to the environment (the higher the score out of 10 the greener the company):

Lenovo:	**8**
Nokia:	**7.3**
Sony/Ericsson:	**7**
Dell:	**7**
Samsung:	**6.3**
Motorola:	**6.3**

Printer cartridges:
Refill your inkjet printer cartridges (www.cartridgesave.co.uk is a good place to start) and use recycled versions wherever possible. For laser printers TonerTopUp (www.tonertopup.co.uk) can supply your business with printer powder that means you can refill, recycle and save money.

One-stop shops:
The following organisations offer advice, inspiration, products and services to help you transform your workplace.

www.greenyouroffice.co.uk
A great company offering eco-friendly cleaning solutions, stationery, IT equipment, food and drink to name a few services.

www.green-office.org.uk
Run by Friends of the Earth Scotland, this great resource offers a quick online audit to see how green your office is (and could be).

www.greenstat.co.uk
Whatever stationery you need they'll have a choice of green options to choose from.

green conferencing

Away-days and team-building days are part of today's workplace culture. The next time you need to organise one, why not suggest one of the following environmentally aware and low-emission locations?

1. **Green & Away (www.greenandaway.org)**
   Europe's only tented conference facility for up to 140 people in rural Worcester (and a favourite of Friends of the Earth) is one of the lowest (if not zero) emission venues available thanks to the 100 per cent wind and solar power electricity supply. They also only use local and organic produce. Be warned: the toilets are composting, which may not suit everyone.

2. **Kindersley Centre (www.sheepdrove.com)**
   Berkshire Downs-based organic farm with stunning state-of-the-art conference centre built using recycled materials (the wash basins are made from old CDs) and packed with the very latest hi-tech gadgetry. Food is local and organic and they can accommodate up to 200 visitors.

3. **The Genesis Centre (www.genesisproject.com)**
   Learn all about sustainable building methods at this purpose-built cathedral of green building in Taunton, Somerset. It's as much of a learning facility as conference centre, which is perfect if you want to combine your work and an interest in new building ideas (possibly the perfect destination for open-minded architects and town planners). The centre holds 200 people and consists of a series of pavilions constructed of earth, straw, clay and timber, with living roofs, and a water pavilion demonstrating the latest water-saving devices.

GREEN TEAM:

Get away from the office in sustainable style at one of the growing number of eco-friendly conference centres like Pines Calyx in Kent.

4. Pines Calyx (www.pinescalyx.co.uk)

The beautiful carbon-neutral earth-sheltered centre in south Kent was built using recycled materials and has been insulated to exceptional standards - not to mention the solar hot-water heating and waste-water recycling. It has room for up to 200 and catering is organic and local as you'd expect.

SCHOOLS

How do you get a school interested in cutting carbon dioxide emissions? What steps as a parent or carer can you take to make a difference? And who can you turn to for help?

Schools may be responsible for less than 2 per cent of UK emissions, but for an individual school there are huge benefits - educationally and financially - in addressing their climate footprint. About four fifths of a school's energy comes from fossil fuels and lighting accounts for around half of electricity costs - already you can see there's huge potential to make savings. What's more, some schools are finding that tackling their carbon footprint provides great educational opportunities and links to the local community.

With the same basic needs as your average house (only on a bigger scale), heating, lighting, cooking and cooling are all essential in the daily running of a school. OK, the gas bill is higher but the principles of energy-saving and reducing carbon dioxide are fundamentally the same; so using less, insulating, changing habits (especially for travel) and microgeneration are still the best ways to slash your child's school's footprint, and reduce the amount of cash that gets spent on fuel - leaving more for useful things like books and teachers.

school buildings

Lighting:

Enormous savings can be made instantly, and with little difference in performance, by simply changing to low-energy bulbs wherever possible. In some instances light fittings may need to be replaced to accommodate the change which would add to the cost, but the energy savings over the years will almost certainly see the bulbs pay for themselves. Schools can also benefit from lighting controls (sensors in the corridor, so lights only come on when someone walks along it) and clearly marked switches so that pupils and staff know which to use to light only the area they need.

Heating:

Making buildings as highly insulated as possible is a major priority. Many old school buildings (like many of our homes) are draughty, so the same principles as detailed in Chapter 7 can be applied. If new buildings are planned, be sure to ask at any consultation meetings what level of insulation will be included, and how the building will make best use of incoming heat and light: does it face south? In a typical school building 10-20 per cent of (heating) energy demand could be met by using the heat from the sun either to generate warmth (solar radiation) or through clever building design that maximises exposure to the sun (passive solar). This could be as much as 40 per cent, not to mention reducing the need for lighting all day.

Even something as simple as suggesting that the heating system has a full service and managing which radiators are turned off will mean that less space is heated unnecessarily. If the budget is available, considering thermostatic radiator valves (see Chapter 7) will also greatly cut the amount of energy wasted in heating. Many old heating systems are horribly inefficient and great savings can also be made by installing two boilers, and using only one in the summer months when demand is less. That said, if your child's school is considering a new heating system, make sure you tell them about the microgeneration possibilities.

Microgeneration:

Because of their size and the way they use energy, schools can often benefit from small-scale on-site energy generation (or microgeneration). Schools tend to have some open space, and a large roof area to make the most of the sun, wind and ground-source heat. The most effective small-scale systems tend to be solar thermal water heaters, but with higher than domestic energy demands schools are also perfectly suited to CHP (combined heat and power) plants that generate heat

top tip

Make sure that the school is using its heat and light as efficiently as possible before considering microgeneration.

and electricity, and biomass boilers that use wood pellets instead of oil or gas to heat the buildings (see Chapter 7). Wind turbines, as you can see from the Brill case study (p.384), can generate carbon dioxide-free power, but, as with homes, location and planning permission is key (although many rural schools often have the advantage of large windy playing fields). Microgeneration isn't a low-budget option: but that's all the more reason substantial school grants and funding should be available. So if you're serious about it, talk to other parents, governors and staff and arm yourself with details of not only what savings can be made, but also the funding that's available. Find out about generating extra revenue (see p.394 for some great ideas).

Making money from microgeneration:
An added advantage to choosing solar or wind technology at school is the potential to generate extra income by selling energy back to the grid. Since schools are closed at weekends and for 13 weeks for holidays it's the perfect opportunity to top up some funds. Renewable energy supplier Good Energy has a Smart Generation scheme set up to make selling back to the grid as simple as possible (www.good-energy.co.uk). But beware: the red tape is notoriously difficult to navigate when dealing with OFCOM – no matter how small your school you'll have the same paperwork as if you were a full-scale power station.

top tip

Controlling energy usage – by making sure it's not too hot or too cold – will improve comfort for staff and students. And comfortable students will be more attentive, making learning more enjoyable for everyone.

what temperature?

Ask what temperature the classrooms are: any higher than the figures below is just a waste.

18°C	Normal teaching
15°C	Circulation spaces (e.g. corridors) and for areas with high levels of activity (e.g. sports halls)
21°C	Special needs schools, low-activity areas or areas with very young children

case study

eco-school

'Having the technology around makes everyone think harder about how much they use, and where it comes from.'

When head teacher Chris Hirst joined Brill Primary School in Buckinghamshire he saw the opportunity to show just what can be achieved with renewable energy.

'When I joined the foundations were there, there was already a solar panel and an organic garden with pond, and it was these encouraging signs that made me look into other ways to turn the school green. I was aware of the renewable technology that existed and was convinced that aside from slashing the running costs of the school, it would be a superb educational experience for everyone involved.

'When I approached the governors with my plan for a wind turbine we quickly realised there was no money for such a scheme and many were sceptical of its benefit. I did my research and realised that there were several grant schemes available including the DTI Clear Skies (which gave £16,000), the TV Energy Community Renewable Energy Grant (£2,000) and the Buckinghamshire County Council Energy Challenge Fund (£8,060). The remaining £6,060 was raised through hard work and determined fundraising by the pupils, parents and teachers.

'We didn't want to push ahead if there was any strong objection so I organised a trip to a local farm that had the same wind turbine as the one I was proposing to install at the school. I soon received the support from the parish council.

'A [turbine] seemed the perfect choice to start with as there had been [one] on the site of the school back in 1600 – now we were bringing it right up to date. The turbine doesn't produce nearly enough power to run the school,

4000 kW is enough for two/three small houses, but it does save £500 on electricity bills each year and 1,720 kg carbon dioxide.

'We've taken it a step further now and have two wind turbines, a ground source heat pump and a solar water heater. Our three newest classrooms are entirely carbon neutral which has been a real achievement.

'It's had a great effect on everyone involved and we now look at everything in a different light, energy saving is second nature and the kids are always looking for new ways to cut the amount they use. All [ages] now deal with these matters and even our year six [10-11year olds] are discussing climate change and how it affects them, what can be done and what gives us adults the right to risk their futures.'

For more information about turning your own school green contact Brill School via their website www.brill.bucks.sch.uk.

fact

An energy audit for one typical Victorian-built primary school in London revealed that for just £100 to fix a dripping tap and fit a hot-water tank jacket, plus a programme of no-cost changes in habits, the school could be more than £2,000 a year better off. And it would save more than 11 tonnes of carbon dioxide a year.

getting involved at school

Give yourself and the school the best chance of cutting emissions and reducing energy use by getting involved. Tutting about decisions made at your child's school that you only find out about in letters sent home won't get you (or the school) anywhere. Being an active voice in the running of the school is easier – and more enjoyable – than you think. Here's how you can play more of an active part.

joining the parent-teacher association (PTA)

Joining the school's PTA will give you the chance to learn about how your child's school operates, and influence in the place where your child spends a huge amount of their waking hours.

Parent-teacher associations (sometimes called the Friends of... your child's school) have no formal powers, but, as with all groups of like-minded individuals, can still get things done. They act as an informal channel of communication between parents and teachers and are often involved in organising fundraising and social activities. Ask your child's teacher for details about your PTA; or contact the National Confederation of Parent Teacher Associations' www.ncpta.org.uk for information and fundraising ideas – in 2006 NCPTA members collectively raised the fantastic sum of £73 million. And don't forget to try and recruit those parents you know who also have an interest in environmental issues (or whatever you're trying to change). The more voices speaking out, the more you will achieve.

becoming a governor

All schools have a governing body that works with the head teacher and senior management team to ensure pupils get a good education. With around 350,000 governor places in England, governors are the largest volunteer force in the country. Parent governor representatives (PGRs) are elected by existing parent governors and it is their job to represent the views of parents to their local authority. The roles of both governors and PGRs are important, and can be excellent ways to find out more about, and influence, education in your child's school or local area.

Some governors are elected by parents, others by local business or faith groups. Think about what you could offer: professional and business skills are valuable, but so is life experience. The job requires commitment – you will attend several meetings each term and these are sometimes held during the day, so you'll need to check that your employer is flexible (although employers are legally obliged to offer some time off).

Working with the school management onside is critical so make sure you involve them from the start. Looking beyond the board is also essential. Think carefully about whom change will affect, who you need on your side to even get a project off the ground. The answer, more often than not, will be the caretaking staff who look after the general day-to-day running. These are the people who understand better than anyone how the school works, how much energy it uses, which classrooms take the longest to heat up, and how often light bulbs need to be changed.

If you are already on the PTA or are a governor, how about setting up a sub-group of individuals to look at ways of cutting carbon. A 'carbon task force' could begin to work out what needs to be done first and make suggestions to the rest of the group.

bags of energy – making the most of your child's enthusiasm

Have you noticed that while grown-ups spend their time debating, questioning and pontificating over the problem of climate change, it's often young people who are actively doing something about it – recycling, raising money, looking after their surrounding environment and saving energy at home and in school? Up and down the country there are hundreds of pupil-run eco-clubs that are often more clued-up and determined than any organised committee of teachers and parents.

These groups should be encouraged, especially if there isn't much sustainability taught in the curriculum. It only takes a couple of motivated pupils and some adult guidance to get an active club up and running and you'll be surprised at how much a small club can achieve – from raising awareness and setting up recycling schemes (collecting cans for charity is especially popular if run as a competition) to basic campaigning in order to get the way their school is run changed for the greener.

changing your family's habits

Even if you're not a governor or on the PTA there are plenty of things you can do to help reduce school-associated emissions.

Walk, cycle, organise a lift share...
do anything you can to reduce the amount you drive to school (see also Chapter 8). You'll reduce the number of road traffic accidents, make the whole family healthier and save up to half a tonne of carbon dioxide each year.

Reinforce good habits.
If it's a bad idea to leave a light on, or the computer on standby at home, why should it be any different at school? Encourage your kids to switch off after they use something and make sure they understand why – that way they can spread the word among their friends.

Recycle.
If the school doesn't provide full recycling, make sure your kids don't simply chuck things away. Provide them with a bag for rubbish with their lunch box so they can bring back their own plastic bottles, cans or packaging where you can dispose of it properly. Better still, ask the school to provide recycling bins.

Make it normal.
To young kids even the strangest things become second nature if they're done often enough. Being environmentally minded is just another way of being responsible, so make it part of the routine and they'll take those values into school with them.

Listen.
The environment is big news on most curricula (and from a very early age). So if your child comes home talking about their green lessons, encourage them to show what they've learnt. You never know, you too might learn something.

GENERATION GAME:

Today's children are tomorrow's leaders. Some are already showing the way to a lower carbon life.

what help is available?

There are an enormous number of people, clubs and organisations out there who can help to reduce the carbon footprint, and raise environmental awareness in schools.

grants and support

Here are some useful organisations to bear in mind.

Carbon Trust (www.carbontrust.co.uk):
A government-funded organisation offering energy-saving advice for schools and businesses. If you're interested in arming yourself with information relating to where, how, and what schools can do to cut carbon, its publications are well worth a browse. Carbon Trust also offers free energy audits for large schools with energy bills over £50,000.
Eco-Schools (www.eco-schools.org.uk):
This campaign coordinated by www.encams.org (the people behind Keep Britain Tidy) was established to encourage schools to work, both with management and in the classroom, towards a more greener, cleaner future. Their aim is to encourage schools to embrace environmentally responsible behaviour.

The Eco-Schools programme can provide a simple framework to help calculate a school's environmental impact and show ways to improve it. Having pupils engaged in the whole process, including monitoring, action planning and decision-making, leads to a genuine feeling of ownership of the programme and an increase in their sense of responsibility for the school environment and local area. To encourage schools and pupils there is an award scheme with Bronze, Silver and Green Flag awards up for grabs.

Building Schools for the Future (www.bsf.gov.uk):
If your local secondary school is looking to build new classrooms or facilities, this government-backed organisation – set up to refurbish or rebuild every secondary school in the country by the end of the decade – is well worth contacting. In 2005-2006 it had £3 billion available in grants for local authorities and schools to spend. The aim is to provide state-of-the-art facilities that save carbon dioxide and also encourage environmental learning at school.
Sustainable Schools (www.suschool.org.uk):
SUSchool exists to help the whole school community learn about sustainability and environmental issues. Run by the Alternative Technology Centre in Hebden Bridge, it offers support, information and advice to schools. The service provides a vast array of teaching aids and lesson planners, games on green themes, kits to help demonstrate principles including solar and wind power, even biomass. Membership costs £50-£250 depending on the size of the school.
Action at School (www.globalactionplan.org.uk):
This scheme, backed by the United Nations Environment Programme, involves experts organising training days that are linked with the curriculum on energy, transport and green issues. Each school forms an Action Team that includes students, teachers, school site managers, caretaking staff and governors, who, through a mixture of games, role-plays and guest speakers, learn how they can make a difference, while a trainer helps them to carry out an audit to pinpoint the best ways to raise awareness and affect change.

shout about it

Each year Friends of the Earth organises a schools activity week. Shout About provides 11- to 13-year-olds the opportunity to explore environmental topics in a fun and engaging way. The organisation provides free curriculum-based activity packs with lesson ideas and visual resources and offers support for schools to hold a whole week of activities, or just a one-off lesson.

Friends of the Earth local groups also offer their time to visit schools and youth clubs. To find a local group in your area, contact Friends of the Earth. The local group network can also help out with resource information, activities, questionnaires and schemes aimed at getting children involved.

When asked why schools should run an activity week about climate change, pupil Jack Allwright says: 'So that children are more aware of what's going on and to show the government and world leaders that we are trying to do something about the future and we want them to help us out.' Visit www.foe.co.uk/learning.

case study

'shout about inspired us to make a difference'

Pupils at the Howard of Effingham School in Surrey got the campaigning and fundraising bug following a week of activities organised by Friends of the Earth – including a rock concert against climate change and an energy audit. Students decided to form their own Youth against Climate Change movement. Richard and Hilary, both in Year 13, explained: 'We have formed this group to raise awareness of this massive and urgent problem. Whilst there are a number of environmental organisations around, there are very few that are run by young people.'

The group designed a website packed with information about climate change and what everyone can do to make a difference. 'We aim for every student in the UK to visit this website within one year,' said Richard. The message of the movement is loud and clear: 'Together this generation will tackle climate change.'

For more visit www.youthagainstclimatechange.org

fundraising

So you've identified the best carbon busters to start your campaign and your applications for grants are being worked on. But you need to start raising some money straight away to pay for things like low-energy light bulbs. Fundraising in schools is a tradition and one that generally win-win – it raises the money and gets everyone involved. There are dozens of ways you can do this, from a dance to a sponsored walk or even a special sports day.

case study
Copthall School

The Copthall School in North London had an annual energy bill of £55,000, and emissions of more than 450 tonnes of carbon dioxide. The school approached the Carbon Trust, which conducted a free site survey. The results showed that Copthall could slash 90 tonnes of carbon dioxide off their footprint and save £36,000 in costs over three years. It was recommended the school set up an energy policy involving pupils and staff, and they now have energy monitors whose job it is to ensure that lights and equipment are turned off when not in use. For more details visit www.carbontrust.co.uk.

oldies but goodies –
classic fundraising events that work

summer fetes, cake sales
discos / barn dances / tea dances...
car boot fairs, jumble sales
prize draws and raffles (canvas local businesses for prizes,in return for advertising... and a good cause)
sponsored walks, swims, runs, skips, jumps, cycles, skating, climbs...
art auctions

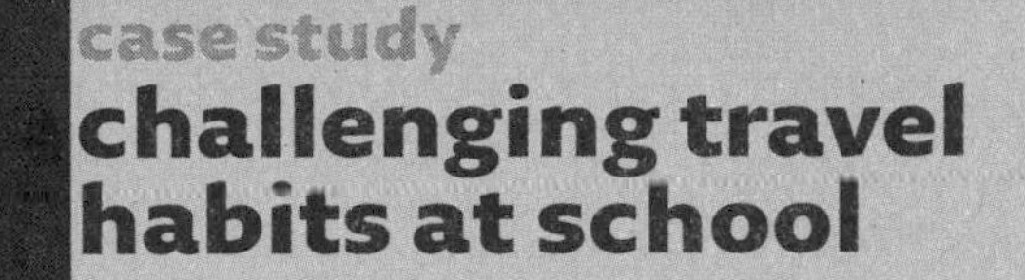

case study

challenging travel habits at school

Students from Mulberry School for Girls, London, challenged teachers to ditch their cars and walk, cycle or use public transport during Shout About week. As an incentive, 50 students offered half a day of admin help to those greening their travel.

The school's energy committee organised activities for the whole school – from green energy debates and after-school cinema screenings to renewable energy demonstrations and a banner with more than 250 pledges.

'It's a fun way of learning about climate change, because instead of just telling everyone about it, we are doing activities which they can be part of,' said 14-year-old Aniqa.

The energy committee – which aims to get young people talking about climate change, promote practical solutions and reduce the school's contribution to climate change – devised an online survey to 'find out what students around the world think about climate change'. Responses came from as far away as India and South Africa, and the students are now hoping to link up with these schools to run joint projects.

For survey results visit www.mulberry.towerhamlets.sch.uk.

could they be carbon detectives?

Set up by the Department for Children, Schools and Families, the Carbon Detectives' Kit is a simple child-friendly calculator designed to help schools explore and measure their own carbon dioxide emissions, identify where the biggest impacts are and plan for improvements. The kit provides advice and info and gives kids the knowledge to put some carbon-cutting into action both in and out of school. For more information visit www.carbondetectives.org.uk.

turn it off

There are around 262 computers in the average secondary school. Leave them on standby and they'll each produce around 12 kg carbon dioxide per month. Switch them all off in the evening and weekends and save the average school 3.14 tonnes of carbon dioxide per month – and that's not counting the printers, scanners or fax machines.

the walking bus

Ease congestion and cut jams by setting up a 'walking bus' to get kids to school safely, and without the car. Also known as a crocodile, the bus involves kids walking in pairs to school along a set route with an adult 'driver' at the front and 'conductor' at the back. The walking bus creates a more formal walking system that allows volunteers to get larger numbers of children to school. Like a bus you can make scheduled stops where children are picked up at specific times but unlike a real bus it is free, healthy and non-polluting.

For a fact sheet on how to set up a walking bus for your neighbourhood, visit www.foe.co.uk.

CARBON IN THE COMMUNITY

We've looked at slashing carbon dioxide emissions at home (see Chapter 7), in schools and in the workplace. But what about the many other facilities we rely on – from drop-in centres, play groups and libraries to sports centres and churches?

It's difficult to find statistics on just how much carbon dioxide – and what percentage of total UK emissions – these community buildings are responsible for. So one of the first things you'll always have to do is an energy audit – i.e. find out how much energy your sports club uses, how much carbon dioxide it's responsible for and how much could be saved. One thing we can be sure of is that, like our homes, schools and businesses, all our community buildings need heat, light and electricity.

If you've switched to energy-saving light bulbs at home, remember the same saving, if not a great deal more, can be made in public and community buildings. The same can be said for virtually all the carbon-saving ideas throughout this book – the big difference is that there are a few more people involved.

simple things you can do to help cut emissions in the community

Before you get embroiled in the world of local politics, fundraising and rabble-rousing, do the easy stuff:

Get some energy saving light bulbs for the hall, club or building. You'll save up to 80 per cent on lighting costs, they'll last for years and save 270 g carbon dioxide per bulb. And you might not even have to pay for them. The Friends of the Earth group in Hammersmith and Fulham discovered it could get free bulbs from the local Energy Efficiency Advice Centre (www.energy-advice.co.uk).

Turn down the thermostat. Just like schools, big old community centres can cost a fortune to heat, but just dropping the temperature by one degree will save up to 10 per cent on heating costs and nobody will notice the difference. And you don't need hot water to be any warmer than 41 degrees, so check the boiler isn't working overtime.

Appoint an Energy Champion. If every club, centre or local group had an Energy Champion who was responsible for encouraging people to think about the energy they used the savings could be enormous. They might be called a nag to begin with, for reminding people to turn electrical items and lights off, but when the energy bills start tumbling people would certainly thank them for it. You might not be the best person for this job unless you have a good knowledge of the building and how it works. It's worth talking to the caretaker, cleaners and grounds staff as they spend the most time in the centre, and almost certainly know everyone who comes in or out. For more details on Energy Champions visit www.everyactioncounts.org.uk.

Hold an awareness session. You can download energy-saving presentations and information from The Community Action for Energy website www.energysavingtrust.org.uk/cafe/. It's a great way to inform people, but also see who's prepared to help out, and how much convincing you'll have to do to make positive changes.

Lag the water tank. Lagging won't cost more than £20 but it'll reduce heat loss from the tank by up to 75 per cent. It'll pay for itself well within the year.

Work out your club's footprint. For a quick calculation of emissions from electricity or gas simply multiply the amount of kWh electricity used (it'll be marked on a bill) by 0.43 and gas (units also shown on the bill) by 0.19 to get the number of kg of carbon dioxide. As a rough guide, an average household is responsible for 10–12 tonnes of carbon dioxide a year. Once you have this information you'll have a starting point to help you see the difference you're making from month to month.

Work out your club's travel emissions. Do a quick survey among members. One person driving a five-mile return trip to attend a meeting will produce roughly 1,500 g of carbon dioxide. Add up all the figures to see your impact and try to discuss arranging lifts, car sharing or using public transport.

Fix leaks. If you have a dripping tap, fix it quickly: in just one day, a dripping hot water tap can waste a lot of energy and enough water to fill a bath.

top tip

If members tend to just roll their eyes when you talk to them about climate change and cutting carbon dioxide, change tack and try the money-saving angle. You'll be amazed at how saving cash can energise the most un-green of people.

top tip

Want to team up with other football fans to tackle your emissions? Check out www.carbonfootyprint.com

case study

'my football team lift-shares'

'I play in a casual football team and up until recently we all drove (with just a couple of exceptions) individually to the pitch each week. This seemed like such a waste of time and energy and also meant that lazy players would always turn up late, so I organised a car pool with each player taking it in turns to pick a car-full up. Once everyone stopped moaning it started to work really well and we now only have three or four instead of nine cars driving each week, plus more of us are now able to enjoy a couple of post-match pints.'

David Gough, Archway (Royston Vasey FC)

who to contact

Need an energy audit, complete carbon overhaul or just a nudge in the right direction? These are the people to ask.

Community Action for Energy (CAfE)

(www.energysavingtrust.org.uk/cafe; 08701 261 444)
This Energy Saving Trust programme is designed to promote local community-based energy-saving projects. If you have a local group (no matter what size) and want to improve your energy efficiency you can sign up for free and get training and advice as well as hearing about funding opportunities.

Every Action Counts

(www.everyactioncounts.org.uk 0845 2410957
This organisation (backed by The Environment Ministry) believes that groups of people can have a bigger impact for a better future. Register your group, club or society on their website and you'll have instant access to expert ideas and teaching materials to get everyone involved and saving energy straight away. They can also help voluntary and community sector workers make the best of their resources and skills.

Community Action Network (CAN)

(www.can-online.org.uk; 0845 456 2537)
CAN provides support for social entrepreneurs and social enterprises (non-profit organisations working with their local community) including reduced-cost office space, strategic support, advice and services to drive your enterprise forward.

Global Action Plan

(www.globalactionplan.org.uk; 020 7405 5633)
This charity aims to get whole streets and neighbourhoods coming together to cut emissions and be responsible for their environmental damage. Your street can become an

Eco Team (six to eight households are recommended per team) and for four months you will measure your waste, gas, electricity, water, transport and shopping habits and then make group changes to lower your impact. The charity also has a range of training materials and aids to help others explain why we should be going green – these include a cycle-powered light-bulb kit that shows just how much easier it is to run an energy-saving bulb.

Envirowise

(www.envirowise.org.uk; 0800 585794)
A free government-funded advice group whose aim it is to get companies of any size to reduce their environmental impact. The website is a great place to pick up tailored advice whatever sector you're in, and it also identifies useful contacts by each region – so if you need to find an energy auditor in Norfolk, say, you can.

Alliance of Religions and Conservation (ARC)

(www.arcworld.org; 01225 758 004)
ARC's main aim is to help the major religions develop environmental programmes based on their own teachings and beliefs. ARC currently works with 11 organised faiths around the world and helps them manage the enormous amounts of land and resources (from church graveyards, housing projects and mosques to entire forests) in an environmentally sound way. Whatever your faith, if you're interested in getting people involved, ARC is well worth talking to.

green volunteering

Whatever your age or abilities there's plenty you can do to help your local community. You'll be surprised at how many environmental projects are going on near you – and all of them could benefit from your help. Here are a few of the best ways to get involved.

Friends of the Earth

(www.foe.co.uk)
Your local Friends of the Earth group will always need support. There are over 220 across the country, all run voluntarlly, all working towards making things better in their area and globally – whether it's campaigning for doorstep recycling, opposing a road scheme or being the public face of a nationwide campaign. Speak to the volunteer coordinator in your local group, let them know of any skills you have that could be useful, and explain how much time you'll be able to spare. And remember, any help will be warmly received because getting things done as a group is much more enjoyable and effective than just the power of one.

Green Gyms

(www2.btcv.org.uk/display/greengym)
Fancy a physical approach to volunteering? Conservation charity BCTV organises Green Gym sessions that combine healthy outdoor exercise with conservation work. Stay fit, save the gym membership, help the planet – what's not to like?

Just do it:

Visit www.do-it.org.uk and search a database of more than 800,000 volunteering positions available across the country. This site can put you in touch with the right people in your area and help match skills, time and interests so you can start making a difference immediately. 020 7250 5700.

Green schemes:

Just type in your postcode and how far you're willing to travel and http://sdvr.sustainable-development.gov.uk/ will bring up a list of voluntary positions available near you. From gardening for the elderly to helping to run your council's recycling demonstrations you'll find plenty of ways to give back and spread a low-impact, low-carbon message.

dos and don'ts

for getting your club or community interested in cutting carbon

do

- Look up Friends of the Earth's website www.foe.co.uk. It's packed full of information on all the key issues, plus practical help on making a difference in your community.
- Think laterally. Who do you know at work, church or college who might have useful contacts?
- Dig around. What community organisations are there in your area? Could you link their interests to climate change? Find out what events they have coming up. Approaching a faith group while they're preparing for a major religious festival might not be the best way to grab their attention – on the other hand, it might be a perfect opportunity to get a lot of people interested.
- Get the facts right. If you're going to talk to a group about

cutting carbon dioxide make sure you know what you want to say. Keep it simple, try and change generic facts to appeal to the interests of the group.

- Have a dummy run. Make sure you know what equipment you'll need and where to get it, and think through any unexpected problems you might run up against on the day. What if the bus is late or if there are more or fewer people than you expected? If possible, plan to give your session with the help of a partner – that way you'll have help with time keeping. Also, two heads are better than one and if one of you is held up or unwell on the day the other person can deliver the session anyway. A dummy run will really boost your confidence before the day.
- Follow up. Be sure to swap contact details and follow up and contact with a reminder of what you're doing, and what they can do. Send a thank you note and make them feel a part of something, it's the best way to forge fruitful relationships.

don't

- Use jargon. If people are looking confused, think about the language you're using. Words like 'lobbying', 'carbon emissions' and even 'climate change' can be confusing. Keep things straightforward and you'll get a better response.
- Make assumptions. Listen to what people say. Part of the value of working with other groups is getting a broader perspective and learning in the process. People are passionate about different things, and if they're going to get active on your campaign it must be on their own terms.
- Expect too much. You can't assume people who are new to an issue or to taking action will immediately be ready to help you (though they might). Signing a petition or committing to receiving more information is a good result you can build on.

case study

microgeneration – power to the people

The village hall in Gamblesby, near Penrith in Cumbria, was in desperate need of renovation and, thanks to the organisation and dedication of a group of just 13 village hall committee members it was able to get funding for a whole range of energy-efficiency improvements, including a ground-source heat pump (GSHP) and a 6 kW wind turbine that now provides carbon-free heating all year round.

The cost of a conventional renovation would have been around £17,500 but the members realised that getting funding for this would be almost impossible in a very competitive funding area.

'We were able to get 83 per cent of the total costs of £42,000 for the pump and insulation, including an £11,000 cheque from the Northern Rock Foundation that really kicked-started everything for us,' explains committee member Bill Mitchell. The other 17 per cent was secured through good old-fashioned community fundraising and support including garden open days, auctions and a unique Gate Festival that raised over £2,000 in one day.

The heat pump has slashed the hall's heating requirements but it wouldn't have been possible without the community pulling together to provide cash and labour (someone had to dig the trenches for the piping). 'There are only around 200 of us in the village, we get on well socially and have a great pool of skill, from farming and handling machinery to IT, finances and PR, so there was always help available and very few arguments, since we were all working towards something we wanted.' The old heaters used 12 kW but the pump now uses just 3 kW and the new 6 kW turbine (again funded thanks to the committee and local support) easily powers it as well as supplying hot water.

'Organising the renovation for the hall has increased my own self-confidence enormously,' explains Bill. 'Since completion I've done twenty-five presentations about what we've been able to achieve and feel empowered – as does the village as a whole. We're even starting to talk about bigger wind turbines, and being a self sufficient village.'

UNLOCK YOUR POTENTIAL:

Will you be among the growing number of community groups tapping into the possibilities that small-scale renewable energy can offer? Check out www.energysavingtrust.org.uk for advice.

bounty hunting: fundraising ideas

Here are some great money-spinning ideas to help fund your energy-saving (or any good cause for that matter) mission.

Street collections:

Sometimes the simplest methods can be the most effective. If you're a member of a local Friends of the Earth group you can get tins, badges and tabards from headquarters to help you raise your profile and money. Seek permission from the local licensing authority – usually the police or county council – before you hit the streets.

Collect stuff:

Mobile phones and inkjet cartridges are both difficult to dispose of, but most of the time they can be reused and there are many companies ready to pay cash for them. Organise a collection box at your club or school and you'll easily be able to generate some funds. Here are some useful contacts: **www.communityfonebank.com** will pay £4 cash for your phone, while www.cellular-surplus.com will actually partner you to help set up a community scheme with the money going to charity.

www.days.org.uk Donate as You Spend will pay 60 pence for each ink cartridge and give the money direct to a charity of your choice (this could be your organisation) while www.inkcycle.co.uk will pay between 5 pence and £5 depending on the type of cartridge. Get collecting and start earning.

De-junk your building:

Go up into the loft of your community centre, village hall, mosque or scout hut and chances are it's packed with under-used junk...and very little insulation. Why not clear out the

HARD NUT TO CRACK?

There's money out there to be had for your carbon-saving scheme if you look carefully.

loft, organise what needs keeping, what can be recycled and then hold a sale with all the profits going towards better loft insulation? You save on bills, clear clutter and cut carbon.

Run a raffle:

Lotteries, raffles and tombolas are great ways to generate interest in a cause, make contact with local businesses and make money. You can approach local companies and ask for prize donations, ask your boss – you'll be surprised what gets given: holidays, booze and vouchers are common. Your raffle will need to be legal, so be sure to check with the Gambling Commission (www.gamblingcommission.gov.uk) first and register if necessary. Most private lotteries and those under £250 won't need to do so.

Organise an auction:

Often highly entertaining evenings, prize auctions can generate considerable sums for charity (especially after dinner when the wine is flowing). As with running raffles you'll need to ask for prize donations, but if you can target companies or friends who can provide a prize someone really wants you may get a lucrative bidding war. If you happen to know someone who's related to someone famous, maybe autographs, pictures or even a round of golf could be organised together as a star prize. If you want to go global, why not open your auction on eBay.co.uk. ?

A car boot stall run by Tunbridge Wells Friends of the Earth not only raised £200 for the group's coffers, it also scored them potential new members and was a great opportunity to get more Big Ask climate change postcards signed.

useful fundraising contacts

www.grantsonline.org.uk
www.fundraising.org.uk
www.charitychoice.co.uk
www.funderfinder.com
www.cafonline.org

BE HEARD

Keen to scale up your effort against climate change? Start here.

If you've read the other chapters in this book you'll have grasped that there are plenty of ways we can beat climate change – and that some of them come down to the things we can each do at home and at work. But there also need to be big changes in the way our energy is produced and used beyond our front door. And that's not something any of us can do alone. For these changes to happen at all - and to happen quickly enough - we need decisive action from our politicians and businesses.

That's where Friends of the Earth comes in. It has an amazing track record of getting politicians and businesses to listen to common sense. From bringing doorstep recycling to every home in England, to persuading the government to introduce the world's first climate change law, Friends of the Earth has huge experience in making people's voices heard

WE WANT A LAW WITH TEETH: **Hundreds of thousands of people joined Friends of the Earth and the Stop Climate Chaos coalition to call on the UK government for tough action on climate change.**

where it counts – in shaping the decisions that affect us all. By joining with like-minded people who are members of Friends of the Earth you can make a bigger difference than on your own.

We're not talking only about the UK or even Europe. Climate change is a global problem that demands global answers. That's why it's so important that Friends of the Earth is an international network: made up of more than 1 million members across every continent: with national organisations in more than 70 countries, it's the most extensive environmental network on the planet. Backed by ordinary people from all walks of life we are able to influence world leaders and governments to make the right decisions.

We've already seen there's a huge amount that could be done to make the climate safer and life better for billions of people. But the systems need to be in place and the price needs to be right for us all to take those opportunities. These sorts of decisions are made not by individuals but by governments. That means as a citizen, resident and voter you are ideally placed to use your influence.

make your mark

Whether you've got the time and inclination to get organising and campaigning right now, would rather dip a toe in the water, or prefer the power of the pen, there are loads of ways you can get involved and have real impact. With Friends of the Earth you can

- make a regular donation to support the organisation's campaigning: more than 90 per cent of its income comes from people giving in this way (see the ad at the back of this book)

PEN PALS:

Signing a petition is one of the easiest ways to get involved and make your voice heard. It works too.

- get active on campaigns that have an impact locally and globally by writing letters, emails and postcards
- become a local campaigner to do more in your area

first steps

It can take just a few canny letters to a retailer, MP, or organisation to get you noticed. Friends of the Earth has won big changes using letter-writing as part of its suite of tactics. While doing these campaigns nationally can involve tens of thousands of letters you might be surprised to learn that you can get an issue raised by your local MP with as few as ten letters; and just 30 on one subject can get it raised at board level at a retailer like M&S.

If you're interested in getting involved in this way a good place to start is Friends of the Earth's website www.foe.co.uk, where you can:

- Sign up to Climate Online, an e-mail newsletter keeping you up to date with the latest news and actions.
- Sign up to Campaign Express – a pack three times a year containing the information you need to send a letter or postcard or make a call to the people that matter. It's an easy way to make a change.
- Sign up to be a Campaign Champion - for people who want to go the extra mile.

case study
the Big Ask

Probably Friends of the Earth's best known campaign to date, the Big Ask launched in 2005 with the support of Radiohead front man Thom Yorke. The campaign's aim was simple but bold: to get the UK government to introduce a law committing it and future governments to year-on-year cuts in carbon dioxide emissions. It would be a world first.

To do this Friends of the Earth had to persuade more than two thirds of UK MPs to support the Bill. That wasn't going to happen behind closed doors. Through an intense

WALK OF LIFE:
The Big Ask campaign helped generate an irresistible movement for change.

“ The best thing about The Big Ask, for me, is the fact that there’s no longer this sense of powerlessness, which is what I had for so long about climate change.

Thom Yorke, Radiohead

media campaign, hundreds of public meetings and street actions, festivals, gigs and film shows around the country Friends of the Earth gave people the chance to ask their MP to back the Bill. Within 18 months, 130,000 people had written, emailed or even visited their MP.

Momentum built. Helping found the Stop Climate Chaos coalition, bringing together major charities and campaign groups, Friends of the Earth also began attracting powerful allies such as The Cooperative Bank, which supported vital research and took the campaign out to millions of its own customers. With celebrities such as Johnny Borrell of the band Razorlight promoting the campaign on his tour, entertainers such as Jude Law and Stephen Fry urging the public to get involved, the call for strong government action had definitely gone mainstream.

The result: in November 2006 the government announced that it would introduce a Climate Bill. Friends of the Earth snapped into action again to ensure that the law was strong enough for the job – the law would be taking its final steps through Parliament around the time this book is published.

En route the Big Ask had picked up a raft of awards and won the support of most MPs across party lines. It would not have worked without the involvement of hundreds of thousands of ordinary people – for many of whom it was their first taste of a political campaign. Announcing the Bill then Prime Minister Tony Blair called it a revolutionary step. The BBC's top political correspondent, Nick Robinson said: 'People often say, "Does anything change politics?" Well it has here.'

top tip

For more on ways of being involved or getting your own local group going contact the Activist Information Officer: 020 7566 1677 localgroups@foe.co.uk

“ Let’s give credit where it’s due. The idea of legally binding CO_2 cuts was first floated by Friends of the Earth. Campaigners drew up a draft bill, pushed the idea relentlessly in The Big Ask, got the signatures of 400 MPs, and were rewarded when the inclusion of a Climate Bill in the Queen’s Speech showed that this was an idea whose time has come.

The Independent, 14 March 2007

case study

‘six months ago, I didn’t even know who my MP was’

When Friends of the Earth needed to lobby every MP in the country to get support for the Climate Change Bill, it turned to its financial supporters. Guy Moore, from Hull, got bitten by the bug. Two months after seeing Al Gore’s climate change film *An Inconvenient Truth* Guy found himself face-to-face with Deputy Prime Minister John Prescott debating the need for a climate change Bill.

The film didn’t just make him realise that he needed to make a personal commitment to reduce his own carbon footprint, but that the Government must play its part too. So when he found out about Friends of the Earth’s effort to lobby every MP, he signed up. ‘Six months ago, I didn’t even know who my MP was,’ says Guy, ‘then, almost overnight, I was organising a meeting with John Prescott, my MP.’

Although Guy reckons he had been well briefed by Friends

of the Earth on how to lobby MPs, he admits: 'I was a bit nervous beforehand, but as soon as I had met him, I realised there was no need to be worried. I even dared challenge him when he claimed that the UK had met its Kyoto targets.'

Like everyone who has been part of The Big Ask campaign, Guy feels proud of its success. He's since been spreading the word in Hull, organising public debates and demonstrations and many more screenings of *An Inconvenient Truth*. 'Getting a climate change bill was a fantastic achievement,' said Guy. 'It's a real landmark.'

620 MPs were lobbied face-to-face in one month in 2006 by Friends of the Earth supporters – that's more than 95 per cent of MPs.

case study

'people wanted to do something but didn't know how'

Undeterred by government foot-dragging, the people of Bishop's Castle in rural Shropshire took it on themselves to cut their carbon emissions.

'A lot of people here are very environmentally conscious. It's not an area full of large, extravagant houses with drives lined with sports cars,' says Alan Terrill, co-ordinator of Bishop's Castle Friends of the Earth. So it was a big surprise when a survey revealed this small rural town on the Shropshire-Wales border to be putting out 40 per cent more carbon dioxide than the national average. 'Many households use coal for heating, as there's no mains gas in the town. Plus, public transport is fairly sparse and there are a lot of old, solid-walled houses,' Alan explains.

Now armed with statistics about the size and source of the problem, the green activists set about decarbonising the town.

'People wanted to do something but didn't know how,' says Alan. 'As we became a focus, more and more people started to drift into meetings.'

Among the campaign's highlights has been the introduction of a fuel pump at the filling station selling biodiesel made from locally sourced, used rapeseed oil. Other low-carbon measures were set to include: an advice service on lowering household emissions, with every interested home receiving a bespoke report and help with grants; car-sharing schemes; cycle paths; local food initiatives; and small-scale and community-scale energy production.

A bonus to the project is the sense of community it has fostered. Bishop's Castle Friends of the Earth used to have nine members: now it has around 100. 'We want this project to act as a catalyst for other areas to feel 'We can do it too',' says Alan. 'Look at our website, copy our ideas – there's no patent.'

Tips and info at www.wasteless.co.uk.

case study

loved up in Manchester

'I love the smug feeling you get as you cycle past all the cars waiting in traffic jams,' says 22-year-old Sarah Whitehead - just one of many commuters tempted on to a saddle by an award-winning ad campaign created by the local Friends of the Earth group in Manchester.

With over £25,000 of Neighbourhood Renewal Fund money, the high-profile campaign on billboards, buses and postcards across the city aimed to get locals to shift from four wheels to two. The group teamed up with 16 local bike shops to offer reductions on gear; and the campaign website (www.loveyourbike.org) with its saucy, loved-up theme, heart-shaped saddles and racy red rear lights, offered loads of information on cycling in Manchester. The website also features an interactive map on which 154 routes and helpful tips have been added by users.

The campaign drew a clear response in favour of better facilities for cycling across the city. 'Before the bike I tended to walk everywhere, unless it was dark or raining, in which case I'd get the bus. Cycling to work is about 20 minutes quicker than walking,' says Sarah, a PhD student at Manchester University. 'I'm really enjoying getting some exercise. I didn't really do any before and I'm glad to have finally found something active that I like.'

Apart from helping cut carbon dioxide emissions, cyclists like Sarah are improving Manchester's air quality and reducing noise pollution. 'I do feel a lot better knowing that I'm doing something for the environment – I'm cycling pretty much every day now.'

After the advertising launch campaigners organised monthly 'critical commute' rides and lobbied for better conditions for cyclists. 'Our message is simple,' said the local group's Graeme Sherriff. 'Manchester City Council has a clear opportunity to make cycling the obvious choice for more people. Should it seize it, the city and its citizens will reap many benefits.'

case study

'I'm part of a big movement'

Within weeks of attending his first Friends of the Earth local group meeting, Richard Reast from Pontypridd met his MP, Kim Howells, and found the right words to convince Kim Howells to raise concerns about climate change with the Prime Minister. Since then Richard reckons he has found a renewed sense of confidence. 'I feel empowered and self-assured now when I talk about climate change. I know I'm part of a big movement making a point that can't be argued with. Best of all the Climate Change Bill is now on the government's to-do list.'

going a step further

Friends of the Earth local groups are run by volunteers in England, Wales and Northern Ireland (Friends of the Earth Scotland is a separate group). There are more than 220 such groups across England, Wales and Northern Ireland and the network is growing. They have an amazing track record of getting things done – in fact some of Friends of the Earth's most outstanding successes are down to them. Because they work locally the groups can talk to MPs and councillors in their area. Decision-makers have to listen to people in their own constituencies, local authorities need to respond to people in their communities, and local businesses pay close attention to what customers on the high street are saying.

You could join a local group tomorrow, start to make a difference – and meet some new people. Why not find out about what's going on in your area? Look out for Friends of the Earth flyers, ads in the local newspapers and on www.foe.co.uk for details of meetings, events and presentations near you. Friends of the Earth groups run hundreds of stalls every year - from your local high street to festivals and farmers' markets.

If joining isn't your thing, there are still plenty of ways you can make your voice heard: for example, by signing up to take part in regular activities involving topics you care about, whether locally or nationally – check out Friends of the Earth's website for options and support.

If there's no local Friends of the Earth group in your area, check if there's one in the next town. Make contact and pick their brains – they're working for the same cause.

Don't forget other community-based groups, too. Different types of organisations – not necessarily green – might be able to help, from providing activists to offering funding. Good groups to approach include: residents' and tenants' associations, schools, clubs, faith groups, your local MP and councillors or local businesses.

why be part of a local group?

Joining a Friends of the Earth local group means you can...

- meet like-minded people and have fun.
- feel more connected to your neighbours.
- be part of changing something you feel strongly about – nothing else feels quite like winning a campaign.
- gain skills in a variety of areas, from public relations and the media to education.
- be at the heart of Friends of the Earth and its national campaigns. If it wasn't for grassroots supporters demanding change, getting crucial legislation such as the Household Recycling Act and Climate Change Law approved would have been much harder.
- get access to expert advice and materials from across the Friends of the Earth network, including its national and regionally based campaigns teams.

request generator

Get information, ask questions and seek answers from those in power with the Friends of the Earth Request Generator. This lets you contact everyone from government departments to water services and local councils. Just type in your details, choose between letter, fax or e-mail, ask the questions you need answering and the website does the rest. For more information visit http://community.foe.co.uk/tools/right_to_know/request_generator.html

find the right person to talk to

Find the right person to talk to in seconds on www.writetothem.com – a fantastic campaigning website. Just type in your postcode and the site brings up a list of your local, district, MP and MEP councillors and contact details for them. It's also packed with useful hints and tips on getting your message across, and what you should expect.

communicate with the council

Often sprawling organisations employing hundreds of people, local councils can sometimes seem impenetrable – but they have the keys to unlock local action on climate change. Here's a quick guide to who you should talk to if you want to get your message heard by your local authority.

Council officers:
Most admin tasks are done by council officers. Officers in charge are often called Directors; look also for names such as Planning Officer or Environment Officer as they will be able to make recommendations to councillors on your behalf.
Ward councillor:
It's these councillors who determine the council's stance on local issues, so it's essential to contact them if you intend to influence policy decisions. They rely on public support to hold on to their positions, so they have a strong reason for taking your concerns seriously and doing something about them.

Other councillors:
You are not limited to contacting only your ward councillor. If you discover another councillor is an avid bird watcher and you're trying to save local trees they may well want to help. Often councillors from other parties want to show how much better a job they could be doing than the party in power, so they might be interested in flexing their political muscle with your campaign.
Committee chairs:
If there is an already-established committee concerned with your issues (and there should be if it's environmental) contact the Chairperson as they'll be influential in the decision-making and can provide you with information on the history of the issue locally.

You'll find the names of all elected officials on your local council's website and local authorities are legally obliged to provide a register of the names and addresses of councillors, plus details of the committees they sit on. The municipal year book will also have this information – you'll find it in your local library.

case study
campaign champions

I'm proud to be a part of it and I'm not a man who joins groups. I'm sick of committees, and this is something I can do on my own or on an ad hoc basis and I hope that like that I can be of service to Friends of the Earth because it is the major lobbying organisation for by far the most important issues facing us. It gives you the opportunity to do something important so if you've half a mind to help the planet this is a way of doing it.

Stephen Whitehead,
Campaign Champion, West Yorkshire

i dare

“Climate change will define this generation. Flitting past us silently is the one chance we may ever have to stand up, look the future in the eye, and say: ‘I dare to make a difference.’ This decisive turn of events will dictate our course for centuries to come: the situation is critical.

Making a difference is about working together, across businesses, governments and, crucially, schools. I plan to tackle energy in my school head on. We want to join with everyone – headmaster, teachers and pupils – to defy fossil fuels, instead embracing wind power. The power to change is infinite. Schools, I believe, can lead the way.

If I could ask people to do just one thing, it would be to think - to sit down and truly reconcile the fact that their actions are exposing nature, themselves and the world’s poorest people to an unprecedented real danger. It is of little use asking people to commit seemingly arbitrary acts of goodness if they are not aware of the implications. True understanding fuels action.

Patrick Kennedy, age 16, is the Young Climate Change Champion for the South East 2008, an award scheme run by the Department for Environment, Food and Rural Affairs (DEFRA)

CHAPTER 10

resources

Back-up info for every chapter

Free information provided by Friends of the Earth may be helpful.
www.foe.co.uk

Friends of the Earth publications are available from **www.foe.co.uk/shop**

Work out your real-world impact and how you can do more:
www.foe.co.uk/calculator

Friends of the Earth Scotland
0131 554 9977
www.foe-scotland.org.uk or **www.green-office.org.uk**

If you like this book you may enjoy a visit to **www.foe.co.uk/living** - which includes advice, articles, interviews and references for *how can i stop climate change?*

Chapter 1 the climate is changing around us

Weather facts: **www.metoffice.gov.uk**
Climate change: **www.metoffice.gov.uk/research/hadleycentre/index.html**
Extreme weather events and the science of climate change: **www.cru.uea.ac.uk**
The science of climate change - a primer:
www.greenfacts.org/en/climate-change-ar4/index.htm
The science of climate change impacts - the global consensus: **www.ipcc-wg2.org**
Exposing the myths about climate change:
environment.newscientist.com/channel/earth/climate-change/dn11638; www.metoffice.gov.uk/corporate/pressoffice/myths/index.html
The risk of floods in the UK: **www.environment-agency.gov.uk/subjects/flood**
Interactive map of the effects of sea-level rise: **flood.firetree.net**
Indicators of climate change in the UK: **www.ecn.ac.uk**
Australia's drought: **www.greenhouse.gov.au/science/hottopics/pubs/topic9.pdf**
Climate change in California: **meteora.ucsd.edu/cap/; www.climatechoices.org**

Climate change impacts in developing countries: **www.christian-aid.org.uk; www.oxfam.org.uk/oxfam_in_action/issues/climate.html**
Desertification: **www.greenfacts.org/en/desertification/; www.cru.uea.ac.uk/tiempo/portal/t26web.htm**
Climate change and wildlife: **www.worldwildlife.org/climate/; www.rspb.org.uk/climate/facts/britishwildlife.asp**

Chapter 2 the science of climate change

Science basics:
www.unfccc.int/essential_background/feeling_the_heat/items/2902.php; www.bbc.co.uk/climate/evidence/
Consensus science – the latest reports from the Intergovernmental Panel on Climate Change: **www.ipcc.ch**
More on the science – Tyndall Centre: **www.tyndall.ac.uk**
Blog by climate scientists: **www.realclimate.org**
Climate change and the UK: **www.ukcip.org.uk**
Science and climate change impacts: **www.bbc.co.uk/climate/impact; www.greenfacts.org/en/climate-change-ar4/**
Climate change and global health: **www.who.int/globalchange/climate/en/**
The economics of climate change – the UK government's Stern Review: **www.hm-treasury.gov.uk./independent_reviews/stern_review_economics_climate_change/stern_review_report.cfm**
The economics of climate change – Lord Adair Turner (first Chair of Committee on Climate Change): **www.sd-commission.org.uk/pages/090207.html**
Climate science and politics: **news.bbc.co.uk/1/hi/sci/tech/portal/climate_change; www.monbiot.com; www.marklynas.org**

Chapter 3 how much carbon can we live with?

Measuring footprints: **www.footprintnetwork.org; www.carbontrust.co.uk**
Interactive map of carbon dioxide emissions: **worldcoolers.org/co2map**
The UN Framework Convention on Climate Change: **www.unfccc.int**
The Kyoto Protocol: **www.unfccc.int/kyoto_protocol/items/2830.php**

Map of 50 US states' greenhouse gas emissions and their national equivalents:
www.sightline.org/maps/maps/Climate-50StEmissions
More about the UK government and climate change (includes links to guides and advice):
www.defra.gov.uk/environment/climatechange
Carbon budgets: **www.foe.co.uk/resource/reports/carbon_budgetting.pdf; www.foe.co.uk/campaigns/climate/news/carbon_budget.html**
The resources we use:
www.neweconomics.org/gen/ z_sys_publicationdetail.aspx?pid=220
More about sustainable food: **www.sustainweb.org**

Chapter 4 saving energy

The Energy Saving Trust on saving energy in the home: **www.est.org.uk**
Building sustainable housing and the Carbon Challenge:
www.englishpartnerships.co.uk/carbonchallenge.htm
Cutting emissions while meeting household energy needs: **www.40percent.org.uk**
The energy-saving house: **www.stoneguardc60researchhouse.co.uk/**
More about eco-renovation and home insulation: **www.ecovation.org.uk**
Energy Performance Certificates: **www.homeinformationpacks.gov.uk/consumer/**
Energy saving in business and the public sector: **www.carbontrust.co.uk**
Cutting carbon emissions in the workplace: **www.globalactionplan.org.uk; www.envirowise.gov.uk**
Carbon offsetting: **www.foe.co.uk/resource/briefing_notes/carbon_offsetting.pdf**
More on the European Union Emissions Trading Scheme:
www.foe.co.uk/resource/consultation_responses/eu_ets_review.pdf
More about the Nottingham Declaration:
www.energysavingtrust.org.uk/housingbuildings/localauthorities/NottinghamDeclaration
Climate change and transport:
www.bettertransport.org.uk/campaigns/ climate_change; www.foe.co.uk/resource/briefings/driving_up_co2_emissions.pdf
Reducing the UK's transport emissions by 2030 - Looking over the horizon:
www.ucl.ac.uk/~ucft696/vibat.html
The Aviation Environment Federation: **www.aef.org.uk**
Climate change and the growth of emissions from UK aviation:
www.foe.co.uk/resource/reports/aviation_tyndall_research.pdf

More about waste and climate change:
www.foe.co.uk/resource/ event_presentations/waste_and_climate_change.pdf
Forests and climate change: **www.forestry.gov.uk/forestry/INFD-6umkar**
Offsetting with tree planting: **www.sinkswatch.org**
The UK government's Energy White Paper (2007):
www.berr.gov.uk/energy/whitepaper/page39534.html
Global detail on mitigating climate change: **www.mnp.nl/ipcc**

Chapter 5 using cleaner energy

Renewable Energy Association: **www.r-p-a.org.uk/home.fcm**
Government guide to renewables: **www.berr.gov.uk/energy/sources/renewables**
More about wind farms: **www.sd-commission.org.uk/pages/wind.html**
British Wind Energy Association: **www.bwea.com**
More about wave power: **www.bwea.com/marine**
More about domestic solar photovoltaic power: **www.solarcentury.com**
More about tidal power: **www.sd-commission.org.uk/pages/tidal.html**
The Severn Barrage: **www.foe.co.uk/resource/briefings/the_severn_barrage.pdf**
Microgeneration for the home: **www.lowcarbonbuildings.org/micro**
More about biofuels: **www.carbontradewatch.org**
Woking's green initiatives: **www.woking.gov.uk/environment/greeninitiatives**
More about an intelligent National Grid: **www.dynamicdemand.org.uk**
Meeting electricity demand: **www.foe.co.uk/resource/reports/bright_future.pdf**
Meeting future energy demand and reducing carbon emissions:
www.foe.co.uk/resource/reports/ living_carbon_budget.pdf;
www.wwf.org.uk/filelibrary/pdf/80percent_report.pdf
More on renewable energy:
www.bbc.co.uk/climate/adaptation/renewable_energy.shtml

Chapter 6 how we can make change happen

More about Dongtan: **www.arup.com/eastasia/project.cfm?pageid=7047**
Pacala and Socolow's wedges (including teaching resources):
www.princeton.edu/ ~cmi/resources/stabwedge.htm
More about contraction and convergence: **www.gci.org.uk**

More about fair shares and cutting carbon: **www.ecoequity.org**
London's Climate Change Action Plan:
www.london.gov.uk/mayor/environment/climate-change
More about St Endellion: **www.stendellion.com**
More about Ashton Hayes: **www.goingcarbonneutral.co.uk**
More about Transition Towns: **transitiontowns.org**
Personal carbon trading: **www.rsacarbonlimited.org/default.aspa**
The Climate Change Levy: **www.cclevy.com**
More on Feed-in Tariffs:
www.foe.co.uk/resource/briefing_notes/feedin_tariff.pdf; www.worldfuturecouncil.org/fileadmin/user_upload/Maja/Feed-in_Tariffs_WFC.pdf
The UK Climate Change Bill:
www.defra.gov.uk/environment/climatechange/uk/legislation/index.htm
More about the Kyoto Protocol: **www.unfccc.int/kyoto_protocol/items/2830.php; news.bbc.co.uk/1/hi/sci/tech/1269921.stm**

Chapter 7 in the home

Calculate your real-world impact with Friends of the Earth's own carbon calculator:
wwww.foe.co.uk/calculator
Government guide to greener living:
www.direct.gov.uk/en/environmentandgreenerliving/greenerlivingaquickguide/
More on energy saving at home: **www.energysavingtrust.org.uk**
More on reducing emissions at home: **www.nef.org.uk/actonCO2/ecohome.htm**
Home improvements at the Energy Saving Trust:
www.energysavingtrust.org.uk/home_improvements
More about green building solutions: **www.greenbuildingstore.co.uk**
For details on all types of insulation: **www.nationalinsulationassociation.org.uk**
More about energy labelling for windows: **www.bfrc.org**
Everything you need to know about your boiler: **www.boilers.org.uk**
Green improvements to an old house: **www.campaignstrategy.org/house/**
Biomass heating:
www.energysavingtrust.org.uk/generate_your_own_energy/types_of_renewables/biomass

How much energy your gadgets and appliances use: **www.energysavingtrust.org.uk/uploads/documents/aboutest/Riseofthemachines.pdf**
For details on the WEEE waste directive: **www.weeeman.org**
Information on the most efficient appliances and electrical goods: **www.sust-it.net**
For an ethical review of a whole host of everyday products: **www.ethiscore.org**
For details of farmers' markets around the UK: **www.farmersmarkets.net**
More about sustainable food: **www.sustainweb.org**
Low-energy building initiatives: **www.bioregional.com**
Home Information Packs: **www.homeinformationpacks.gov.uk/consumer**
More about the Hockerton housing project: **www.hockertonhousingproject.org.uk**
More on the eco-building architects Zed: **www.zedfactory.com**
More on the Earthship Brighton eco homes project: **www.lowcarbon.co.uk**
Find out what green properties are on the market: **www.greenmoves.com**
More on microgeneration (Centre for Alternative Technology): **www.cat.org.uk**
More on microgeneration (including grants): **www.lowcarbonbuildings.org.uk/micro**
For more information on solar power: **www.solartradeassociation.org.uk**
More about wind power from the British Wind Energy Association: **www.bwea.com**

Chapter 8 out and about

The Environmental Transport Association: **www.eta.co.uk**
Government advice on greener travel: **www.direct.gov.uk/en/environmentandgreenerliving/greenertravel**
Find out just how efficient your car is: **www.vcacarfueldata.org.uk**
More on driving economically: **www.ecodrive.org**
More on reducing carbon emissions behind the wheel: **www.dft.gov.uk/actonCO2**
Guidance on buying a greener car: **www.whatgreencar.com**
A network of car and taxi sharing, walking and cycle groups: **www.liftshare.org**
Details and map of UK car clubs: **www.carplus.org.uk/carclubs/what-are.htm**
More on aviation growth's problems and solutions: **www.airportwatch.org.uk**
Travelling without flying: **www.noflytravel.com**
For more information on cycling around the UK: **www.sustrans.org.uk**
Discount cycle equipment via the Cycle To Work scheme: **www.cyclescheme.co.uk**
More about the benefits of cycling: **www.cyclingengland.co.uk**
Help with setting up your own school cycle run: **www.saferoutestoschools.org.uk**
Compare public transport options with a car route: **www.transportdirect.info**
Train travel in the UK, Europe and beyond: **www.seat61.com**

Chapter 9 scaling up your impact

More info on turning your work place green:
www.carbontrust.co.uk/energy/assessyourorganisation
More advice and help on greening your office: **www.envirowise.gov.uk**
How organisations can reduce emissions: **www.nef.org.uk/greencompany/index.htm; www.globalactionplan.org.uk**
Online audit of your office's green credentials: **www.green-office.org.uk**
More on the Parent Teachers Association: **www.ncpta.org.uk**
More on sustainable schools projects: **www.eco-schools.org.uk; www.suschool.org.uk**
Online carbon footprint calculator for English schools in England:
www.carbondetectives.org.uk
Brill Church of England Combined School's eco project: **www.brill.bucks.sch.uk**
Find out more about turning your schools green: **www.foe.co.uk/learning**
Get support, information and ideas for your community projects:
www.everyactioncounts.co.uk;
www.energysavingtrust.org.uk/cafe
Find out who needs volunteers in your area: **www.do-it.org.uk**
Friends of the Earth's campaigns: **www.foe.co.uk/campaigns**
Friends of the Earth's information request generator:
community.foe.co.uk/tools/right_to_know/request_generator.html
Find the right person to lobby: **www.writetothem.com**

Find out more ...

Paul Brown, *Global Warning: The Last Chance For Change* (Guardian Books 2006)

Lester R Brown, *Plan B2.0: Rescuing A Planet Under Stress And A Civilization In Trouble* (Norton & Co, the Earth Policy Institute 2006)

Chris Goodall, *How to Live a Low-Carbon Life* (Earthscan 2007)

Al Gore, *An Inconvenient Truth: The Planetary Emergency Of Global Warming And What We Can Do About It* (Bloomsbury 2006)

Tony Juniper, *How Many Lightbulbs Does It Take to Change the Planet? – 95 Ways to Save Planet Earth* (Quercus 2007)

Tony Juniper, *Saving Planet Earth* (Collins 2007)

Mark Lynas, *High Tide* (Harper Perennial 2005)

Mark Lynas, *Six Degrees: Our Future on a Hotter Planet* (Fourth Estate 2007)

Mark Lynas, *Carbon Calculator: Easy Ways to Reduce Your Carbon Footpint* (Collins 2007)

George Marshall, *Carbon Detox: Your Step by Step Guide to Getting Real About Climate Change* (Gaia Thinking 2007)

George Monbiot, *Heat: How to Stop the Planet Burning* (Allen Lane 2006)

Fred Pearce, *The Last Generation: How Nature Will Take Her Revenge for Climate Change* (Eden Project Books 2006)

Sue Roaf, *Ecohouse* (Architectural Press 2007)

Joseph J Romm, *Hell & High Water: Global Warming: The Solution And The Politics And What We Should Do* (William Morrow 2007)

Andrea Smith & Nicola Baird, *Save Cash and Save the Planet* (Collins 2005)

index

Endangered species...